Mathematical Investigations

AN INTRODUCTION TO ALGEBRAIC THINKING

DeMarois • McGowen • Whitkanack

Student Support Manual

Ernest East
Northwestern Michigan College

Phil DeMarois
William Rainey Harper College

Mercedes McGowen
William Rainey Harper College

Darlene Whitkanack
Indiana University

▲▼ ADDISON-WESLEY

An imprint of Addison Wesley Longman, Inc.

Reading, Massachusetts • Menlo Park, California • New York • Harlow, England
Don Mills, Ontario • Sydney • Mexico City • Madrid • Amsterdam

Reproduced by Addison-Wesley Educational Publishers Inc. from camera-ready copy supplied by the authors.

Copyright © 1998 Addison-Wesley Educational Publishers Inc.

All rights reserved. No part of this publication may be reproduced, stored in a retrieval system, or transmitted, in any form or by any means, electronic, mechanical, photocopying, recording, or otherwise, without the prior written permission of the publisher. Printed in the United States of America.

ISBN 0-321-01050-7

1 2 3 4 5 6 7 8 9 10 VG 00999897

CONTENTS

SECTION ANSWERS 1

Chapter 1 3

Section 1.1	Learning Mathematics	3
Section 1.2	Thinking Mathematically	4
Section 1.3	Using Variables to Generalize	5
Section 1.4	Expanding the Notion of Variable	7
Section 1.5	Review	10

Chapter 2 12

Section 2.1	Whole Number Domains	12
Section 2.2	Order of Operations with Whole Numbers	15
Section 2.3	Algebraic Extensions of the Whole Numbers	17
Section 2.4	Properties That Change the Order of Operations	20
Section 2.5	Review	22

Chapter 3 24

Section 3.1	Investigating Relationships Numerically	24
Section 3.2	Function: Algebraic Representation	28
Section 3.3	Function: Notation for Algebraic Representations	34
Section 3.4	Function: Geometric Interpretations	36
Section 3.5	Triangular Numbers	40
Section 3.6	Power and Factorial Functions	42
Section 3.7	Review	43

Chapter 4 49

Section 4.1	Integers and the Algebraic Extension	49
Section 4.2	Operations on Integers	51
Section 4.3	The Absolute Value Function	56
Section 4.4	Graphing with Integers	59
Section 4.5	Using a Graphing Utility	63
Section 4.6	Functions over the Integers	69
Section 4.7	Review	75

Chapter 5 — 80

- Section 5.1 Rates of Change — 80
- Section 5.2 Rational Numbers and Proportional Reasoning — 86
- Section 5.3 Investigating Rational Number Operations — 90
- Section 5.4 Reciprocal and Power Functions — 94
- Section 5.5 Integer Exponents — 98
- Section 5.6 Review — 100

Chapter 6 — 104

- Section 6.1 Real Numbers and the Algebraic Extension — 104
- Section 6.2 The Square Root Function — 106
- Section 6.3 Classes of Basic Functions — 110
- Section 6.4 Linear Functions — 113
- Section 6.5 Quadratic Functions — 121
- Section 6.6 Review — 124

Chapter 7 — 130

- Section 7.1 Linear Equations and Inequalities in One Variable — 130
- Section 7.2 Systems of Equations — 137
- Section 7.3 Finding Zeros of Quadratic Functions by Factoring — 143
- Section 7.4 Additional Factoring Experiences — 149
- Section 7.5 Review — 154

TI-82 GRAPHING CALCULATOR REFERENCE MANUAL — 164

TI-82 Reference Manual — 166

TI-82 Index of Procedures — 217

TI-83 GRAPHING CALCULATOR REFERENCE MANUAL — 220

TI-83 Reference Manual — 222

TI-83 Index of Procedures — 275

Section

Answers

CHAPTER 1

What is Mathematics?

Section 1.1 Learning Mathematics

p. 11

1. Opinion

2. Some patterns include

   a. Column 1 begins at twenty and decreases by ones and, if extended, would end at zero.

   b. Column 2 begins at eighty and decreases by fours and, if extended, would end at zero.

   c. Column 3 begins at zero and increases by ones and, if extended, would end at twenty.

   d. Column 4 begins at zero and increases by twos and, if extended, would end at forty.

   e. Column 5 is a constant of twenty.

   f. Column 6 begins at eighty and decreases by twos and, if extended, would end at forty.

3. In the equation $4c + 2(20 - c) = 66$, the four is the number of wheels on a car, the c is a variable representing the number of cars, so $4c$ is the number of wheels on c cars. The two is the number of wheels on a motorcycle, the $20 - c$ is a variable expression representing the number of motorcycles, so $2(20 - c)$ is the number of wheels on the motorcycles. The 66 represents the total number of wheels in the parking lot.

4. $2m + 4(20 - m) = 66$ where m is the number of motorcycles.

5. $c + m = 20$ represents the fact that the sum of the number of cars and the number of motorcycles equals twenty.
   $4c + 2m = 66$ represents the fact that c cars have $4c$ wheels; m motorcycles have $2m$ wheels, and the total number of wheels is sixty–six.

6. The Bulls made twenty–three two–point shots and four three–point shots. This result could be found by guess–and–test, a table, the equation $2t + 3(27 - t) = 58$, where t represents the number of two–point shots, or the system $\begin{array}{l} t + s = 27 \\ 2t + 3s = 58 \end{array}$ where t is the number of two–points shots and s is the number of three–point shots.

Section 1.2 Thinking Mathematically

p. 25

3. a. $SP \xrightarrow{\text{Rule 1}} SPP \xrightarrow{\text{Rule 1}} SPPPP \xrightarrow{\text{Rule 3}} SCP$

   b. $SP \xrightarrow{\text{Rule 1}} SPP \xrightarrow{\text{Rule 1}} SPPPP \xrightarrow{\text{Rule 1}} SPPPPPPPP$
   $\xrightarrow{\text{Rule 3}} SPPPCPP \xrightarrow{\text{Rule 2}} SPPPCPPC$

   c. $SP \xrightarrow{\text{Rule 1}} SPP \xrightarrow{\text{Rule 1}} SPPPP \xrightarrow{\text{Rule 1}} SPPPPPPPP$
   $\xrightarrow{\text{Rule 2}} SPPPPPPPC \xrightarrow{\text{Rule 3}} SPPPPCC$

4. a. S P C S is not a valid path since it contains the letter S in a position other than the first position.

   b. $SP \xrightarrow{\text{Rule 1}} SPP \xrightarrow{\text{Rule 1}} SPPPP \xrightarrow{\text{Rule 1}} SPPPPPPPP$
   $\xrightarrow{\text{Rule 1}} SPPPPPPPPPPPPPPP \xrightarrow{\text{Rule 2}}$
   $SPPPPPPPPPPPPPPPC \xrightarrow{\text{Rule 3}}$
   $SPPPPPCPPPPPPPC \xrightarrow{\text{Rule 3}} SPPPPPCPPPPCPC$

   c. S P P P C is not a valid path since it contains 3 P's. A valid path cannot have a P–count that is a multiple of three.

   d. C S P P is not a valid path since all valid paths must begin with the letter S.

   e. S P P C P P is a valid path. Derive the path. It's challenging but doable.

   f. S C C C C C is not a valid path since it contains 0 P's and 0 is a multiple of three. The P–count cannot be a multiple of 3.

5. A test for non–valid paths might be to check if the path does not begin with the letter S or if it has a P–count that is a multiple of 3.

6. Rules 1 and 2 lengthen valid paths; Rules 3 and 4 shorten valid paths.

7. When working backwards, you think about what rule might have been used to obtain the given valid path. An illegal reversal would mean doing a rule in reverse. For instance, if you did the reverse of what Rule 3 says to do you would replace a C with three P's. This is not a valid procedure, however, and would not be guaranteed to produce a valid path.

8. 0 is a multiple of 3 since there is a whole number, 0, that we can multiply 3 by to obtain 0.

9. The new path is S P P P P P P P P. The number of elements after the S must be doubled. Doubling means multiply the number of elements by 2; counting by 2 means "add 2".

10. SC is not a valid path in the SPC system. Since the number of P's in SC is 0 and 0 is a multiple of 3 it is impossible to arrive at this path by starting with SP and using the rules. This uses the conclusion that a path is not valid if the P-count is a multiple of 3. See the answer to Exploration 5, above.

11. Answers vary. But, in general, when you applied one of the rules to produce a new path you worked within the system.

12. Answers vary.

13. Answer varies. But, drawing general conclusions about the effect of the rules on the P-count of a path as in Exploration 6 above represents working outside the system.

14. Opinion.

15. Opinion.

16. Opinion.

Section 1.3 Using Variables to Generalize

p. 40

3. Using Rule 1, draw a box around the sequence of letters that can be represented by the variable in each of these paths.

   a. S⎡P⎤   b. S⎡C P⎤

   c. S⎡P P P P C⎤P P P P C   d. S⎡C P P C⎤

4. Using Rule 2, draw a box around the sequence of letters that can be represented by the variable in each of these paths.

   a. S P: empty   b. S⎡C⎤P

   c. S⎡P P P P C P P P⎤P   d. S C P P C: Rule 2 cannot be applied.

5. Using Rule 3, draw a box and a circle around the sequence of letters that can be represented by the variables in each of these paths.

   a. S[P C]P P P(P) is one answer.   b. S P P P(P C P) is one answer; box is empty.

   c. S[P]P P P : circle is empty.   d. S[P]P P P(C C P C P P) is one answer.

6. Using Rule 4, draw a box and a circle around the paths that can be represented by the variables in each of these paths.

   a. S[P]C C(P)   b. S C C(P P C) : box is empty.

   c. S[P P P]P P P P: Rule 4 doesn't apply.

   d. S P P P C C(P C P P)

7. a. 1 x 7 = 7 for example.

   b. $7 \cdot 7 \cdot 7 = 7^3$ for example.

   c. $\sqrt{2^2 + 3^2} \leq 2 + 3$ for example.

8. a. Paper is an example of a noun.   b. Boss is a noun that ends in s.

   c. Ax is a noun that ends in x.   d. Fold is a verb.

9. a. Let the variable x represent the singular form of the noun. Then xs is the plural form.

   b. Let the variable x represent the singular form of the noun that ends in s up to the letter s. Then $x$ses is the plural form.

   c. Let the variable y represent the singular form of the noun that ends in x up to the letter x. Then $y$xes is the plural form.

   d. Let the variable x represent the present tense of a verb. Then $x$ed is the past tense.

10. English grammar rules don't work as nicely as the rules in the SPC system because many grammar rules require one to be aware of exceptions.

11. a. $1^2 = 1$; $11^2 = 121$; $111^2 = 12321$.

    b. $1111^2 = 1234321$; $11111^2 = 123454321$.

    c. $111111^2 = 12345654321$; $1111111^2 = 1234567654321$;

    e. 111111111111^2 doesn't follow the previous pattern due to the carry that will occur.

12. John conjectured, based on two examples, that he could park illegally without receiving a ticket. The probable implication of this generalization is he will most likely receive a ticket in the near future.

13. The assumptions about variables with examples are:
In a problem or statement, every time a variable is used, it represents the same thing.

    Example: In $2x + 3x$, the x represents the same number in both places.

    The variable probably represents something different when we work a new problem.

    Example:

    | Rule 1 | S x $\longrightarrow$ S x x |
    | Rule 2 | S x P $\longrightarrow$ S x P C |
    | Rule 3 | S x P P P y $\longrightarrow$ S x C y |
    | Rule 4 | S x C C y $\longrightarrow$ S x y |

    The variable x means something different in the statement of each rule.

    Variables make it possible to describe a few specific examples as a generalized statement.

    Example: $n + 0 = n$ is an example of generalizing the fact that the sum of any number and zero is the number.

    A generalized statement must always be true and, in order to guarantee this, the use of restrictions, or conditions, may be required.

    Example: $\sqrt{x}$ represents a real number only if x is not negative.

Section 1.4 Expanding the Notion of Variable

p. 57

2. Techniques that are algebraic include writing an equation or writing a system of equations. Both these techniques make use of variables to represent and solve the problem.

3. y is the base and m is the exponent.

4. When you square a number larger than one, you always get a number larger than the original number.

5. When you square a number between zero and one, the square is always smaller than the original number.

6. a. For $2n$, we are computing the product of two and the variable n. For n^2, we are computing the product of n and n.

    b. For 2^n, we are computing the product of n twos.

7. The larger the value of p, the smaller the value of $50 - p$.

8. We will use the formula $P\text{-}count = 2^n - 3x$ where n is the number of times Rule 1 is applied and x is the number of times Rule 3 is applied.

    a. In this case, $n = 7$ and $x = 11$.
        So $P\text{-}count = 2^7 - 3(11) = 128 - 3(11) = 128 - 33 = 95$.
        There are 95 P's in the resulting valid path.

    b. In this case, $n = 4$ and $P\text{-}count = 7$. So, $7 = 2^4 - 3x$.
        $7 = 16 - 3x$. Since $7 - 16 = -9$, Rule 3 must have been applied three times.

    c. In this case, $x = 6$ and $P\text{-}count = 46$. So, $46 = 2^n - 3(6)$.
        $46 = 2^n - 18$. Since $46 = 64 - 18$, we know $2^n = 64$ which leads to $n = 6$. Rule 1 was applied six times.

    d. If Rule 1 is applied four times, then the $P\text{-}count$ is $2^4 = 16$. We could then apply Rule 3 either 0, 1, 2, 3, 4, or 5 times.

    e. By applying Rule 3 successively, we get possible $P\text{-}counts$ of 16, 13, 10, 7, 4, and 1.

9. a. To obtain a $P\text{-}count$ of 13, I must apply Rule 1 four times and Rule 3 once. See the answer to 8d.

    b. To obtain an $P\text{-}count$ of 52, I must apply Rule 1 six times to obtain 64 P's. then eliminate twelve P's. This requires four applications of Rule 3.

    c. Answers may vary. Most likely, the answer was found numerically by an informed trial–and–error process using a calculator.

10. In Section 1.3 the term domain is used in two different ways. On page 30 it refers to the possible paths to which a rule in the SPC system may be applied. SCPPCPP is a path in the domain of Rule 2, for instance. But in using variables to state the rules on page 36 of the text x represents possible sequences of letters that could occur between S and P when applying Rule 2 to a valid path. In SCPPCPP for instance $x =$ CPPCP. The term domain is used to refer to all the possible sequences of letters that might appear between S and P in a valid path where Rule 2 could be applied. In other places in Section 1.3 and Section 1.4 variables represent numbers in algebraic expressions and equations. Depending on the use of the expression in a problem situation the possible numbers which could replace the variable may be restricted. In the P-count equation after Rule 1 has been applied to the path SP six times Rule 3 could be applied 0, 1, 2, 3, 4, or 5 times only. The domain of x in

8

this situation is this set of numbers. The domain of a variable is the set of all possible replacement values of that variable. The domain of one of the rules in the SPC system is the set of all possible paths to which a rule may be applied.

11. Rewrite each expression using only one exponent, if possible.

   a. $(7^5)(7^{11}) = 7^{16}$
   b. $(2^9)^5 = 2^{45}$
   c. not possible
   d. $(2^9)(2^5) = 2^{14}$
   e. not possible
   f. $(9^{15})(9) = 9^{16}$
   g. $(11^{23})(11^4) = 11^{27}$
   h. $(5)(5^2) = 5^3$
   i. not possible
   j. $(11^{23})^4 = 11^{92}$

12. The difference is in the order of the operations which are indicated. $(2^9)^5$ means raise 2 to the 9th power first and then raise the number you get to the 5th power. $2^9 + 2^5$ means to compute the exponents first and then add the results while $(2^9)(2^5)$ means to compute the exponents first and then multiply the results.

13. $(n)(n)$ means to multiply n by itself and $n + n$ means to add n to itself.

14. Write each as an expression with one exponent.

   a. $(y^6)(y^4) = y^{10}$
   b. $y^5 \cdot y = y^6$
   c. $y^m y^n = y^{m+n}$
   d. $(y^6)^4 = y^{24}$
   e. $(y^m)^n = y^{mn}$
   f. $(y^4)^6 = y^{24}$

15. If Rule 1 is applied six times, the number of P's in the resulting path is $2^6 = 64$. Then reduce the number of P's by three. Recall that x represents the number of times Rule 3 is applied. The domain of the variable is the set of all possible values of that variable. Rule 3 can be applied no times or until less than three P's remain. So, the domain of the variable x in the P–count equation is $\{0, 1, 2, 3, 4, 5, 6, 7, 8, 9, 10, 11, 12, 13, 14, 15, 16, 17, 18, 19, 20, 21\}$. If Rule 3 is applied twenty–one times, 63 P's are removed and one P remains.

16. Variables were used in mathematical expressions to represent numbers from the domain of the problem situation. This contrasts with the use of variable in Section 1.3 where variables were used to represent paths that remained constant when a rule was applied. Also, variables were used in this section to generalize a pattern.

17. a. $k + k$ or $2k$ since Rule 1 doubles the number of characters after the S.

b. $k + 1$ since Rule 2 adds one character, C, to the path.

c. $k - 2$ since Rule 3 replaces three P's with C. This is a decrease of 2 characters so we subtract 2 from the original number of characters.

d. $k - 2$ since Rule 4 omits two C's. This is a decrease of 2 characters so we subtract 2 from the original number of characters.

Section 1.5 Review

p. 63

1. a. One could visualize the process with a diagram or picture. Adding or multiplying could be used to come up with the answer of 9 minutes. This, of course, ignores the time required to move from one cut to another.

   b. One must determine the types of coins. Constructing a table with the headings pennies, nickels, dimes, and quarters and filling in possible entries which would total to 43 cents would be one approach. Applying logic one might reason that there could be only 3 pennies. One could soon see that he had no quarters.

   c. Constructing a table to organize the information and search through the possibilities would be a good approach. The table could have headings like number of stools, number of legs on the stools, number of tables, number of legs on the tables, sum of the number of legs. Another approach might be to set up the equation $3s + 4t = 31$ and explore the possible whole number solutions to this equation. In the equation s would stand for the number of stools and t would be the number of tables the carpenter made. There are three possible solutions: 1 stool and 7 tables, 5 stools and 4 tables, or 9 stools and 1 table.

   d. Insight into the problem is the key to this one. Perhaps you should imagine working backwards. Consider when the basket would be full and how much of the basket would be filled 1 second before that time.

   e. A calculator would be useful to produce the number of minutes early you could leave work each day on this one. It would also be great if you could detect a pattern and see how to calculate the solution. You might end up trying to solve the equation $2^{z-1} = 8 \cdot 60$, where the left side represents how many minutes early you would be able to leave on day x. Find x by entering this expression in the Y= menu on your calculator and examining a table that is produced from this expression.

2. a. $a \cdot b = b \cdot a$ b. $\dfrac{0}{a} = 0$ as long as a is not 0.

   $5 \times 10 = 10 \times 5$ $\dfrac{0}{50} = 0$

c. $0 + a = a$ d. $1 \times a = a$

$0 + 23 = 23$ $1 \times 7 = 7$

3. $SCCPPPPC \xrightarrow{Rule\ 4} SPPPPC \xrightarrow{Rule\ 3} SPCC \xrightarrow{Rule\ 4} SP$

4. Rewrite each of the following using only one exponent, if possible.

   a. $(5^{19})(5^8) = 5^{27}$ b. $(3^5)^4 = 3^{20}$

   c. $(3^5)(3^4) = 3^9$ d. not possible

   e. $(17)(17^5) = 17^6$ f. not possible

   g. not possible   h. $(19)(19) = 19^2$

   i. not possible

5. Rewrite using only one exponent, if possible.

   a. $(b^3)(b^{11}) = b^{14}$ b. not possible

   c. not possible   d. $(k^8)(k) = k^9$

   e. $(z^5)^3 = z^{15}$ f. $(b^2)^9 = b^{18}$

   g. $(x)(x) = x^2$ h. not possible

   i. To check part e for instance we replace z with 2 and compute the original expression and our simplification using the specified order of the operations.

   $(z^5)^3 = (2^5)^3 = (32)^3 = 32,768$ and $z^{15} = 2^{15} = 32,768$

6. The order of operations is different in each of the expressions. $(3^5)^4$ indicates raise 3 to the fifth power first then raise the number you get to the fourth power. $(3^5)(3^4)$ would be computed first by doing the two exponents and then multiplying the numbers you get while $3^5 + 3^4$ would be computed by first doing the exponents and then adding the answers.

7. The first indicates multiplication and the second expression indicates addition.

8. a. Negative numbers are not used to measure lengths of sides of rectangles.

   b. No one has probably lived that long.

c. It has been known to get that cold in Minnesota in January.

d. There are teachers who have made that salary.

e. There aren't that many hours in a week.

9. Opinion

10. Opinion

CHAPTER 2

Whole Numbers: Introducing a Mathematical System

Section 2.1 Whole Number Domains

p. 83

3. a. {1, 3, 5, 7, 9, 11, ...}

   b. {0, 2, 4, 6, 8, 10, ...}

   c. {0, 1, 2, 3, 4, 5, 6, 7, 8, 9}

   d. {0, 1}

   e. {2, 3, 5, 7, 11, 13, 17, 19, 23, 29, 31, 37, ...}

   f. { } or ∅

   g. {4, 6, 8, 9, 10, 12, 14, 15, 16, 18, 20, 21, 22, 24, 25, 26, 27, 28, 30, 32, ...}

   h. {0, 1}

4. a. The set of odd numbers consists of the whole numbers that are not divisible evenly by 2.

   b. The set of even numbers consists of the whole numbers that are divisible evenly by 2.

   c. The set of digits is the whole numbers less than 10.

   d. The set of whole numbers that are their own squares is the whole numbers that are less than 2.

   e. The set of prime numbers consists of the whole numbers that have no divisors except 1 and themselves.

f. The set of whole numbers between 11 and 12 is empty since there are no whole numbers between 11 and 12.

g. The set of composite numbers consists of the whole numbers larger than 1 that have more than two distinct divisors.

h. The set of whole numbers that are neither prime nor composite consist of the whole numbers less than 2.

5. a, b, e, and g are infinite since they have no greatest element; c, d, f, and h are finite since the number of elements in the set is a finite number.

6.   a. Not closed under addition; closed under multiplication.

    b. Closed under addition; closed under multiplication.

    c. Not closed under addition; not closed under multiplication.

    d. Not closed under addition; closed under multiplication.

    e. Not closed under addition; not closed under multiplication.

    f. A moot point!

    g. Not closed under addition; closed under multiplication.

    h. Not closed under addition; closed under multiplication.

7. Prime factorizations

    a. $2^2 \cdot 61$ b. $2 \cdot 3^3$

    c. $3^3 \cdot 2^2$ d. $2 \cdot 5 \cdot 41$

    e. $13 \cdot 11^2$ f. $3 \cdot 7 \cdot 11 \cdot 17$

    g. $2^3 \cdot 5^2 \cdot 7^2$ h. $5^2 \cdot 13^2 \cdot 17$

8. No, because 51 isn't prime.

9. It should be $2^3 \cdot 3 \cdot 5$.

10. The largest element of the set is 80. The next whole number is 81, which is used as the first number above the upper limit of the set. Non–whole numbers between 80 and 81, such as 80.36578, could have been used. The answer also could be stated as the set of whole number multiples of four that are less than **or equal to** 80.

11. 64 represents the starting number of P's. By applying Rule 3, three P's are removed. The division of 64 by three is used to determine how many times three P's can be removed.

12. a. {S, P, C}; finite

    b. {S P}; finite

    c. {Rule 1, Rule 2, Rule 3, Rule 4}; finite

    d. The set of sequences containing only the letters S, P, and C; infinite

    e. The set of strings produced by applying a rule to the initial path or to a previously derived valid path; infinite

    f. The empty set Ø; finite

13. The set of valid paths is a subset of the set of paths. Every valid path is a path but there are many paths that are not valid paths.

14. Yes. If one applies a rule to a valid path in the SPC system you create another valid rule. So one cannot leave the set of valid paths by applying one of the rules.

15. a. This is the set of whole numbers less than 2.

    b. This is the set of one–digit, odd whole numbers.

    c. This is the set of squares of whole numbers.

    d. This is the set of whole number multiples of 7.

16. Zero is not a prime number since it has an infinite number of divisors. One is not a prime number since it has exactly one divisor. Prime numbers have exactly two divisors.

17. a. 287 is not prime since 287 is divisible by 7 and 41.

    b. 283 is prime. It is not divisible by the primes 2, 3, 5, 7, 11, and 13. No larger prime is a divisor of 283 since the square of all other primes exceeds 283.

    c. The largest prime that needs to be tested to determine if 283 or 287 is prime is 13. If a number has a divisor other than itself and one, the divisor must be a prime whose square is less than the given number.

18. a. $m + 9$

    b. $m - 4$

    c. $2m$

    d. $2m + 3$

e. $m + 1$ and $m + 2$

19. a. The set of all whole numbers that are greater than or equal to 4. $\{4, 5, 6, ...\}$

    b. The set of all whole numbers.

    c. The set of all even whole numbers. $\{0, 2, 4, 6, ...\}$

20. Answers are limited only by your imagination. An example might be to determine how many feet there are in 40 inches.

21. The symbol for the empty set is $\emptyset$. An example of an empty set is the set of labrador retrievers registered for this class.

Section 2.2 Order of Operations with Whole Numbers

p. 101

3. a. $5(7 + 6) = 65$
    b. $18 - 2(3 + 4) = 4$

    c. $9 - 3(2) = 3$
    d. $\dfrac{8}{9-5} = 2$

    e. $28 + 10 \div 5 = 30$
    f. $17 - 3 + 5 + 8 - 2 = 25$

    g. $11(5) - 3(4) = 43$
    h. $(8 + 10) \div 4 + 3(2) = 10.5$

    i. $\dfrac{10 + 8}{5 + 4} = 2$

4.

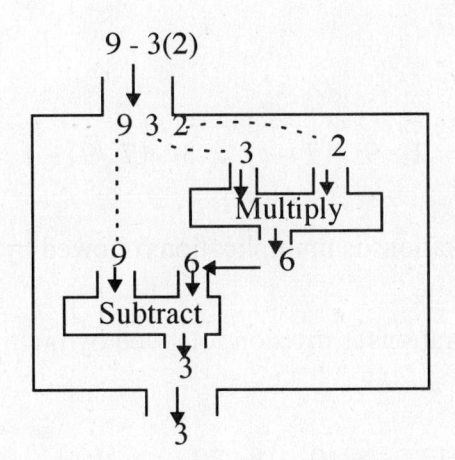

5. Rewrite each of the following using at most one exponent, if possible.

    a. $(7^0)(7^{11}) = 7^{11}$
    b. $(2^9)^0 = 1$

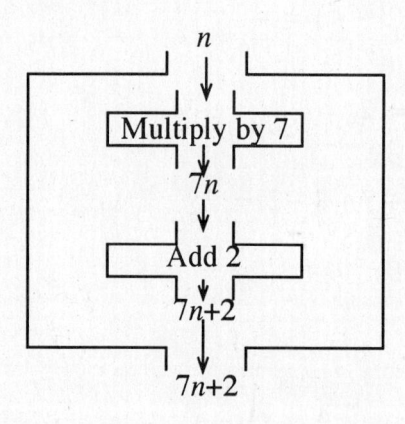

c. $2^9 + 2^0 = 2^9 + 1$ d. $(2^9)(2^0) = 2^9$

e. $(5^3)(0^5) = 0$ f. $(9^0)(9) = 9$

g. $(4^{23})(4^4) = 4^{27}$ h. $(7)(7^9) = 7^{10}$

i. $8 + 8^2 = 8(1 + 8) = 8(9)$ j. $(4^0)^4 = 1^4 = 1$

6. $\dfrac{5 + 2(4^3 - 6)}{\sqrt{8+1}+7} - 2 = \dfrac{5 + 2(64 - 6)}{\sqrt{9}+7} - 2 = \dfrac{5 + 2(58)}{3+7} - 2 = \dfrac{5+116}{10} - 2 = \dfrac{121}{10} - 2$

   $= 12.1 - 2 = 10.1$

7. a. $n + 3$ b. $5n$

   c. $n - 8$ d. $7n + 2$

8.

   n → Multiply by 7 → $7n$ → Add 2 → $7n+2$ → $7n+2$

9. a. 9   b. 30

   c. -2 d. 44

10. Some examples: $9 - 7 - 1 = 1$; $9 - 7(1) = 2$; $9 - (7 - 1) = 3$; $(7\sqrt{9}) - 1 = 20$.

11. $\dfrac{20}{2 \cdot 3} = \dfrac{20}{6} = \dfrac{10}{3} = 3\dfrac{1}{3}$. The order of operations is multiplication followed by division.

    $20/2 \cdot 3 = 10(3) = 30$. The order of operations is division followed by multiplication.

12. An example: Start with 11.

    11 → 34 → 17 → 52 → 26 → 13 → 40 → 20 → 10 → 5 → 16 → 8 → 4 → 2 → 1

Section 2.3 Algebraic Extensions of the Whole Numbers

p. 112

2.   a. $2(5) = 10$ b. $5^2 = 25$

    c. $5 + 11 = 16$ d. $6 - 5 = 1$

    e. $3 + 2(5) = 13$ f. $4(5) - 7 = 13$

    g. $5^2 + 3 = 28$ h. $3(5)^2 - 4 = 3(25) - 4 = 75 - 4 = 71$

    i. $(5)^2 + 6(5) = 55$ j. $6 + 7(5) + (5)^2 = 66$

    k. $9(5) - 2(5)^2 + 8 = 45 - 50 + 8 = 3$

3.   a. e. Multiply, add; h. Square, multiply, subtract; j. Square, multiply, add, add.

    b. e. Multiply 2 and 5. Add 3 to the product.

        h. Square 5. Multiply the square of 5 by 3. Subtract the 4 from the product.

        j. Square 5. Multiply 7 by 5. Add 6 to the product of 7 and 5. Add the square of 5 to this sum.

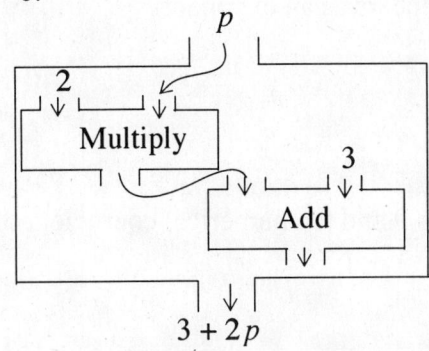

    c. e.

        [Diagram: Inputs 2 and p go into "Multiply"; its output and 3 go into "Add"; final output is $3 + 2p$.]

17

h.

```
         p
         ↓
  ┌─────────────────────────┐
  │  ┌──────┐      ┌─3──┐   │
  │  │Square│      │Multiply│ │
  │  └──┬───┘      └──┬─┘   │
  │     └──→ ┌─4───┐ ←┘     │
  │          │Subtract│     │
  │          └───┬─┘        │
  └──────────────┼──────────┘
                 ↓
             $3p^2 - 4$
```

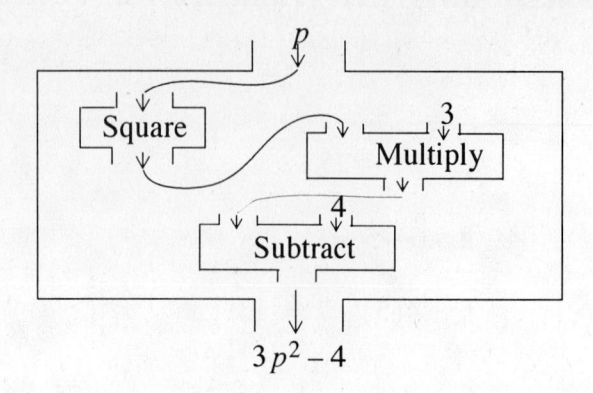

j.

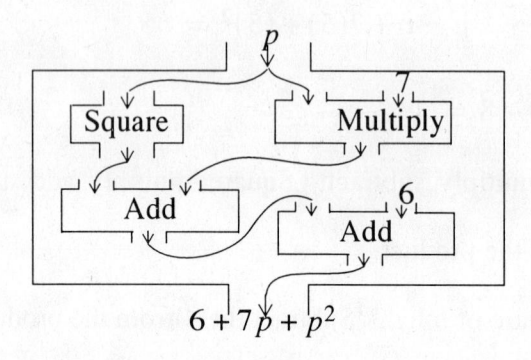

```
          p
          ↓
  ┌────────────────────────────┐
  │ ┌──────┐        ┌──7──┐    │
  │ │Square│        │Multiply│  │
  │ └──┬───┘        └──┬──┘    │
  │  ┌─┴──┐         ┌──┴──6─┐  │
  │  │ Add│         │  Add  │  │
  │  └─┬──┘         └───┬───┘  │
  └────┼────────────────┼──────┘
                ↓
          $6 + 7p + p^2$
```

4. $\frac{1}{2}n^2 + \frac{1}{2}n$

5. This is not a polynomial since the variable n appears as an exponent.

6. In the linear polynomial $9x + 7$

   a. The terms are $9x$ and 7.

   b. The numerical coefficient of the first term is 9 and the numerical coefficient of the second term is 7.

7. In the quadratic polynomial $4x^2 + x + 3$

   a. the terms are $4x^2$, x, and 3.

   b. the numerical coefficients are 4, 1, and 3.

8. Write an expression with at most one exponent for each of the following expressions. List any restrictions on the variables.

   a. $(y^6)(y) = y^7$ b. $(y^6)^0 = 1$, y cannot be 0

c. $(y^0)^6 = 1$, y cannot be 0   d. $(y^4)(y) = y^5$

e. $(y^m)(y^0) = y^m$, y cannot be 0   f. $(y^m)^0 = 1$, y cannot be 0

9. a. Output 1 is 21; Output 2 is 29.

   b. A1 is 49; A2 is 14; A3 is 35; Output is 40.

10. a. $3x + 8$ b. $x^2 - 2x + 5$

11. a.

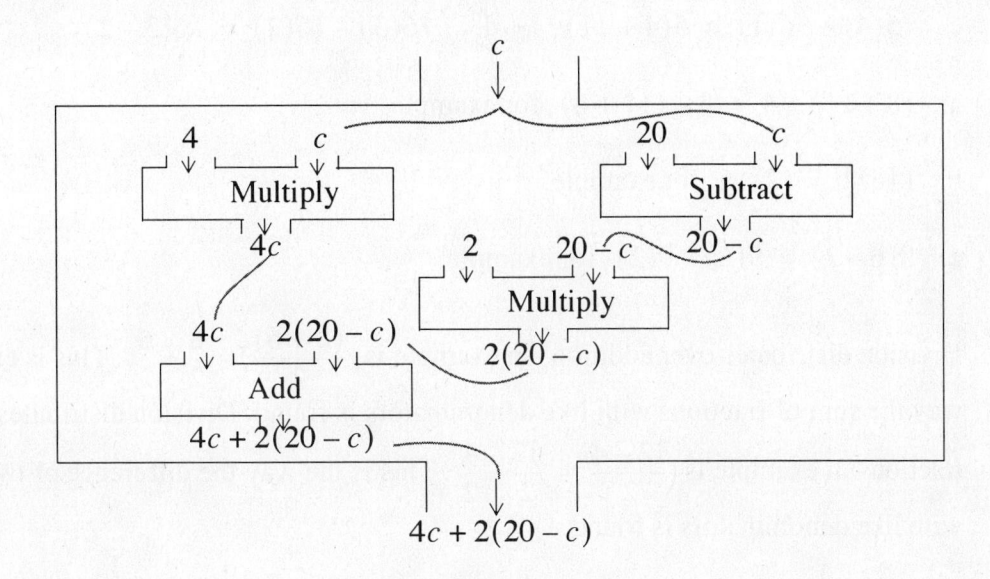

b.

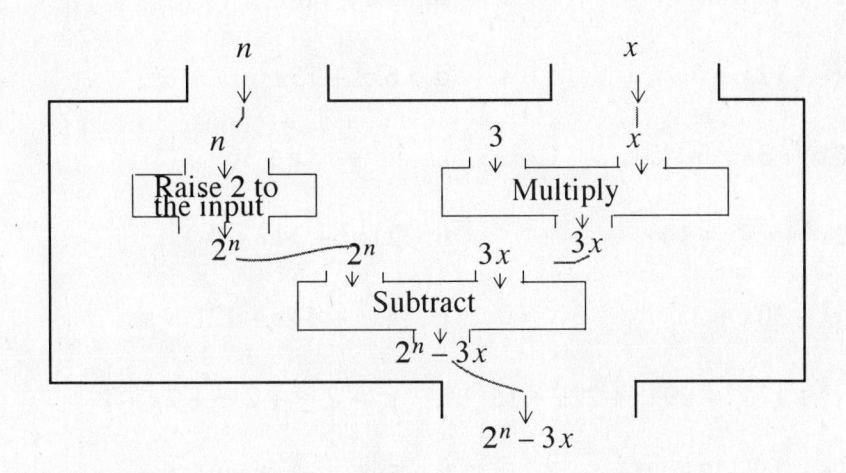

12. a. For example: Input 3; Output 10. Input 15; Output 10. Input 92; Output 10.

    b. $[(2n + 9) + n] \div 3 + 7 - n$

19

Section 2.4 Properties That Change the Order of Operations

p.132

2.   a. $18 + 9 = 9 + 18$ b. $7(11) = 11(7)$

3.   a. $5 + (3 + 7) = 5 + 3) + 7$ b. $(5(3))7 = 5(3(7))$

4.   a. $7(4 + 9) = 7(4) + 7(9)$ b. $4(8 - 3) = 4(8) - 4(3)$

      c. $5(4) + 5(11) = 5(4 + 11)$ d. $16(3) - 16(2) = 16(3 - 2)$

5.   a. $(8 + 11) + 4 = 8 + (11 + 4)$, for example.

      b. $(16)2 = (2)16$, for example.

      c. $9(6 - 2) = 9(6) - 9(2)$, for example.

6.   Division distributes over addition–an example is $\frac{(8 + 6)}{2} = \frac{8}{2} + \frac{6}{2}$. This is exactly the way the sum of fractions with like denominators is found. Division distributes over subtraction–an example is $\frac{12 - 4}{2} = \frac{12}{2} - \frac{4}{2}$. This is the way the difference of two fractions with like denominators is found.

7.   a. $3x^2 + 4x + 9$ b. $5x + 11$

      c. $5x^2 + 12x + 5$ d. $6x^3 + 15x^2$

      e. $10x^3 + 35x^2 + 45x$ f. $x^2 + 13x + 36$

      g. $12x^2 + 29x + 14$ h. $21x^2 + 34x + 4$

      i. $9x^2 + 30x + 25$ j. $4x^2 + 44x + 121$

      k. $2x^4 + 17x^3 + 29x^2 + 22x + 15$ l. $x^4 + 2x^3 + 2x^2 + 2x + 1$

8.   a. $7(x + 2) = 7x + 14$ b. $5(8 - m) = 40 - 5m$

      c. $6b - 18 = 6(b - 3)$ d. $35 + 7t = 7(5 + t)$

9.   a. $9x^2 + 18x = 9x(x + 2)$ b. $21x^4 + 14x^3 = 7x^3(3x + 2)$

      c. $5x^3 + 25x^2 + 30x = 5x(x^2 + 5x + 6) = 5x(x + 3)(x + 2)$

d. $3x^2 + 6x + 8$ is prime.

10. a. $(3 \cdot 5)^2 = 3^2 \cdot 5^2$ b. $(3+5)^2 = 3^2 + 2 \cdot 3 \cdot 5 + 5^2$

     c. $(ab)^2 = a^2 \cdot b^2$ d. $(a+b)^2 = a^2 + 2ab + b^2$

11. a. The associative property of addition allows you to rewrite the sum.

$$9 + 6 = 9 + (1 + 5) = (9 + 1) + 5 = 10 + 5$$

     b. $9 + 8 = 10 + 7 = 17$

     c. $99 + 57 = 100 + 56 = 156$

12. 9 could be replaced with $10 - 1$ and then the multiplication by 367 could be distributed over this difference. You would need to compute $3670 - 367$ which could be done mentally to get 3303.

13. a. $5a - 8 = 12$ b. $7z + 5 = 12$
        $5a = 20$ $7z = 7$
         $a = 4$ $z = 1$

14. a. Rule 1: Sx → Sx x

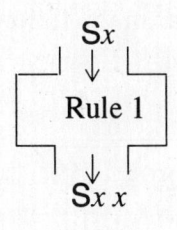

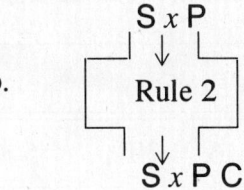

     b. Rule 2: S x P → S x P C

     c. Rule 3: S x P P Py → S x Cy

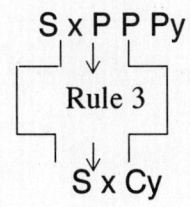

     d. Rule 4: S x C C y → S x y

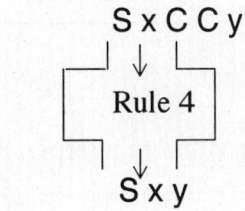

Section 2.5 Review

p. 136

1. Yes.

2. No. 71^2 is the prime factorization.

3. Your response.

4. $\dfrac{7-2(1^3+2)}{\sqrt{3(7)+4}} - 7 = \dfrac{7-2(3)}{\sqrt{21+4}} - 7 = \dfrac{7-6}{\sqrt{25}} - 7 = \dfrac{1}{5} - 7 = -6\dfrac{4}{5}$

5. $5 \cdot 3 - 2 \cdot 3^2 + 3 = 15 - 18 + 3 = 0$

6. 4, 8, 6, 12, 24, 166, 332, 664, 249, and 1992

7. Answers will vary. Check with others in your class.

8. a. $3^2 \cdot 7 \cdot 11$ b. $2^3 \cdot 5 \cdot 13^2$

   c. $29 \cdot 31$ d. $3^4 \cdot 7 \cdot 11 \cdot 17$

   e. $7^2 \cdot 47$ f. 8191 since it's prime.

9. To generalize means to draw conclusions about a class of objects or about the way a set of operations works based on observations of only a few (or many) of the objects of a few instances where the property is satisfied. For instance, if we observe that multiplication over division seems to work in the three examples which appear in the following table,

$a(b+c)$	$a \cdot b + a \cdot c$
$5(2+4) = 5 \cdot 6 = 30$	$5 \cdot 2 + 5 \cdot 4 = 10 + 20 = 30$
$4(1+8) = 4 \cdot 9 = 36$	$4 \cdot 1 + 4 \cdot 8 = 4 + 32 = 36$
$8(2+7) = 8 \cdot 9 = 72$	$8 \cdot 2 + 8 \cdot 7 = 16 + 56 = 72$

   The generalization is that $a(b+c) = ab + ac$ for all numbers a, b and c.

10. Note: answers may vary due to interpretation of problem.

    a. Only whole numbers are appropriate since this represents a count.

    b. Whole numbers are not appropriate since measurements involve fractional amounts and rounding.

c. This could involve numbers other than whole numbers since age can be measured in fractions of time units.

d. Only whole numbers are appropriate since this represents a count.

e. This could involve numbers other than whole numbers if the cost of a can of pop is not a multiple of ten cents.

f. Only whole numbers are appropriate since this represents a count.

11. a. $(7^{23})(7^2) = 7^{25}$ b. $(5^7)^2 = 5^{14}$

    c. $(5^7)(5^2) = 5^9$ d. $5^7 + 25$ or $25(5^5 + 1)$

    e. $(11)(11^0) = 11$ f. $(11)(5^0) = 11$

    g. $(2^0)(3^2) = 3^2$ h. $(2^3)(0^2) = 0$

12. a. $(b^5)(b^9) = b^{14}$ b. $s^7 + a^0 = a^7 + 1$ as long as a is not 0

    c. $(z^0)(a^6) = a^6$ d. $(x^4)x = x^5$

    e. $(a^0)^3 = 1$ f. $(b^2)^0 = 1$

    g. $x^0 x^0 = 1$ h. $x^0 + x^0 = 1 + 1$

13. a. $n + 5$ b. $3n$

    c. $4 - n$ d. $3n + 7$

14. a. 14   b. 27

    c. -5   d. 34

15. $3x - 2 = 3 \cdot 5 - 2 = 13$

16. a. $18x + 52$ b. $9x^2 + 4x + 8$

    c. $21x^3 + 24x^2$ d. $10x^2 + 53x + 63$

    e. $64x^2 + 80x + 25$ f. $35x^3 + 66x^2 + 37x + 6$

17. a. $4(4x + 3)$ b. $2x^3(2x^2 + x + 9)$

    c. $14x^2 + 27x + 7$ is prime.   d. $3(5x^2 + 8x + 1)$

CHAPTER 3

Functional Relationships

Section 3.1 Investigating Relationships Numerically

p. 156

2. Every function is a relation. Not all relations are functions. Since every function is a relation and functions have only one output for each input relations don't always have multiple outputs.

3. {0,1,2,3,...}; Infinite; it has no last element.

4. {0, 1, 2, 3, 4, 5, 6, 7}; Finite; it has only 8 elements.

5. {0, 1, 2, 3, 4, 5, 6, 7}; Finite; it has only 8 elements.

6. {1, 2, 3, 4, 5}; Finite; it has only 5 elements.

7. {0, 1, 2, 3, ...}; Infinite; it has no last element.

8. {1, 2, 3, 4, 5}; Finite; it has only 5 elements.

9. {0, 1}

10. A function is a process that receives input and returns an unique value called the output. A relation is a process that receives input and can return more than one unique output for a given input. All functions are relations but not all relations are functions.

11. a. $1 + 3 + 5 + 7 + 9 = 25$

    $1 + 3 + 5 + 7 + 9 + 11 = 36$

    $1 + 3 + 5 + 7 + 9 + 11 + 13 = 49$

    $1 + 3 + 5 + 7 + 9 + 11 + 13 + 15 = 64$

    b. The sum of the first n odd whole numbers is equal to the nth square.

c.

Sum of *n* odd numbers	Number of odd numbers (*n*)
1	1
4	2
9	3
16	4
25	5

Table 1: Sum of *n* odd numbers

d. 1 + 3 + 5 + 7 → **Count elements** → 4

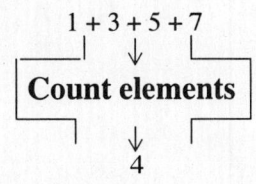

e.

Number of odd numbers (*n*)	Sum of *n* odd numbers
1	1
2	4
3	9
4	16
5	25

Table 2: Number of odd numbers (*n*)

f. 4 → 1 + 3 + 5 + 7 → 16

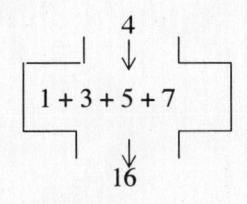

g. 1 + 3 + 5 + 7 → **Count elements** → 4 → 1 + 3 + 5 + 7 → 16

25

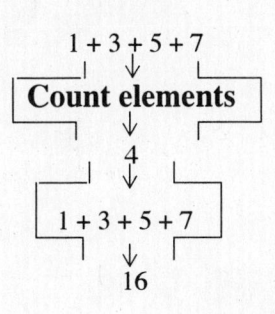

h. The output of the expression $2n - 1$ is the nth odd counting number. So, the sixth odd counting number is $2 \cdot 6 - 1 = 12 - 1 = 11$. The sum of the first six odd counting numbers is $6^2 = 36$. The sum of the first n odd counting numbers is given by the value of n^2.

i. In each problem we determined two functions that were put together so that the output of the first function was the input of the second function. Notice how the function machine diagram in Figure 8 (page 155) and the function machine diagram that appears above in the solution to part g illustrate this connection.

12. a.

Number of cars	Number of cycles
0	20
5	15
10	10
15	5
20	0

Table 3: Cars and cycles

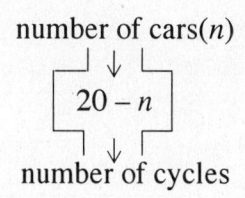

number of cars(n)
↓
$20 - n$
↓
number of cycles

b.

Number of cars	Total wheels
0	40
5	50
10	60
15	70
20	80

Table 4: Cars and wheels

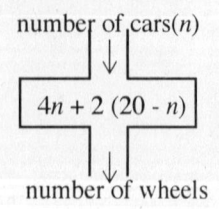

number of cars(n)
↓
$4n + 2(20 - n)$
↓
number of wheels

13. a. Each row begins and ends with a 1.

6	15	20	15	6		
7	21	35	35	21	7	
8	28	56	70	56	28	8

   b. The row sums are 1, 2, 4, 8, 16, 32, 64, 128, 256. The sum of the *n*th row is the *n*th power of 2.

   c. The non-zero whole numbers 1, 2, 3, 4, 5, 6, ... appear by reading down two different diagonals; outside diagonals are always ones; triangle is symmetric.

   d. Yes, whole numbers play an important role. The squares seem to be involved as well as adding consecutive whole numbers. Notice the numbers 1, 3, 6, 10, ... are found by adding 1, 1 + 2, 1 + 2 + 3, etc. This sequence of numbers appears in the table too.

14. This problem is so much fun and takes time to investigate. We do not want to spoil your fun by suggesting an answer yet. Keep exploring.

15. Stop when you produce a valid path already produced.

16. a. [Rule 1: input S*x*, output S*x x*] — a function; a given input produces an unique output.

    b. [Rule 2: input S*x* P, output S*x* P C] — a function; a given input produces an unique output.

    c. [Rule 3: input S x P P P y, output S x C y] — a relation; more than one output possible for a given input.

d.
```
  S x C C y
      ↓
   [Rule 4]
      ↓
    S x y
```
a relation; more than one output possible for a given input.

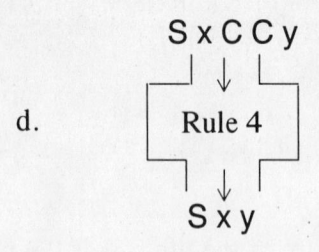

17. We offer a few suggestions for approaching this problem. First, construct a table like the one on page 144 to familiarize yourself with the problem. Second, look at the finite differences to see if any number seems to be important. Does a remainder seem to provide you with a good predictor?

Section 3.2 Function: Algebraic Representation

p. 171

2. a. 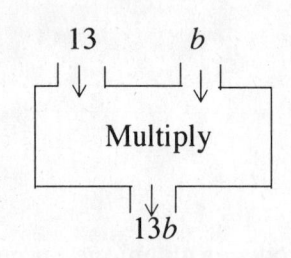
```
    13      b
      ↓    ↓
   [Multiply]
        ↓
       13b
```

b.
```
          z
          ↓
      [Cube]
        ↓ z³
         z³      4
           ↓    ↓
         [Multiply]
              ↓ 4z³
            4z³
```

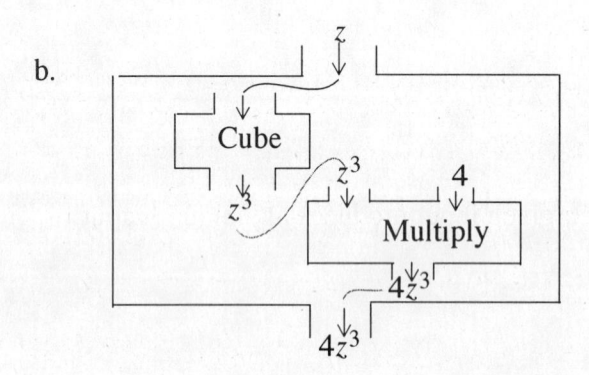

c. 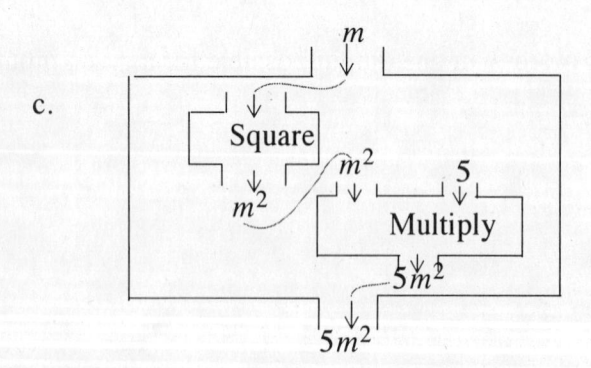
```
          m
          ↓
      [Square]
        ↓ m²
         m²       5
           ↓    ↓
         [Multiply]
              ↓ 5m²
            5m²
```

d.

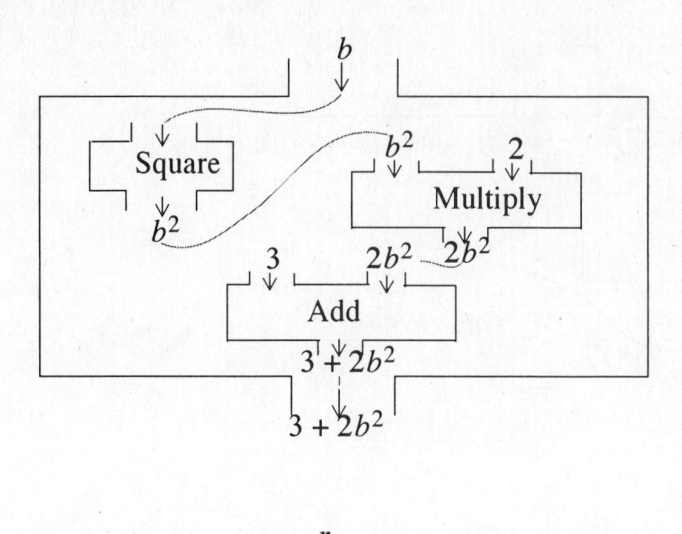

Input: b

Square → b^2
Multiply: b^2, 2 → $2b^2$
Add: 3, $2b^2$ → $3 + 2b^2$

Output: $3 + 2b^2$

e.

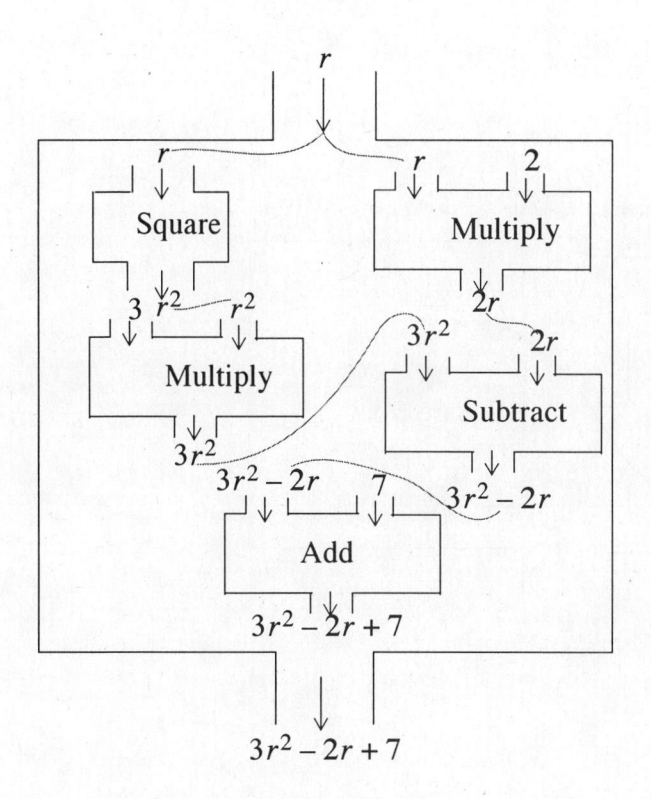

Input: r

Square → r^2
Multiply: r, 2 → $2r$
Multiply: 3, r^2 → $3r^2$
Subtract: $3r^2$, $2r$ → $3r^2 - 2r$
Add: $3r^2 - 2r$, 7 → $3r^2 - 2r + 7$

Output: $3r^2 - 2r + 7$

3. a.

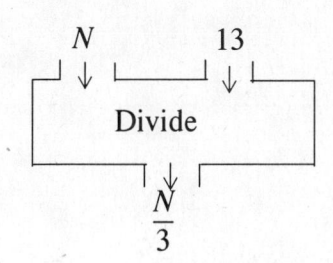

Divide: N, 13 → $\dfrac{N}{3}$

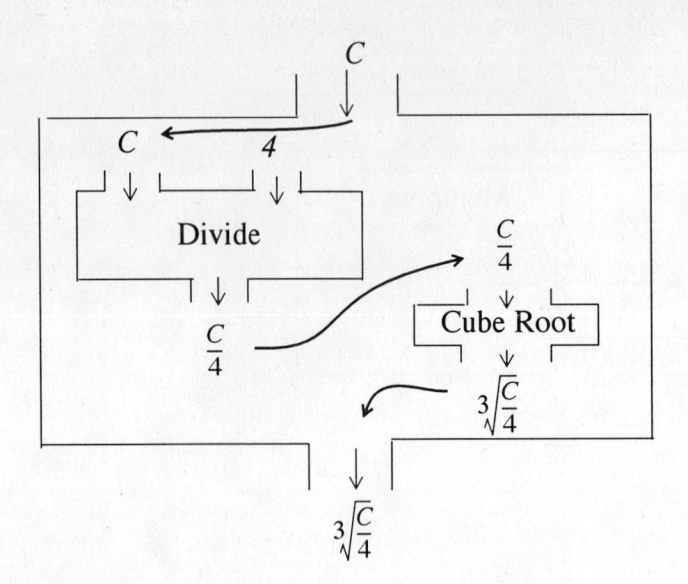

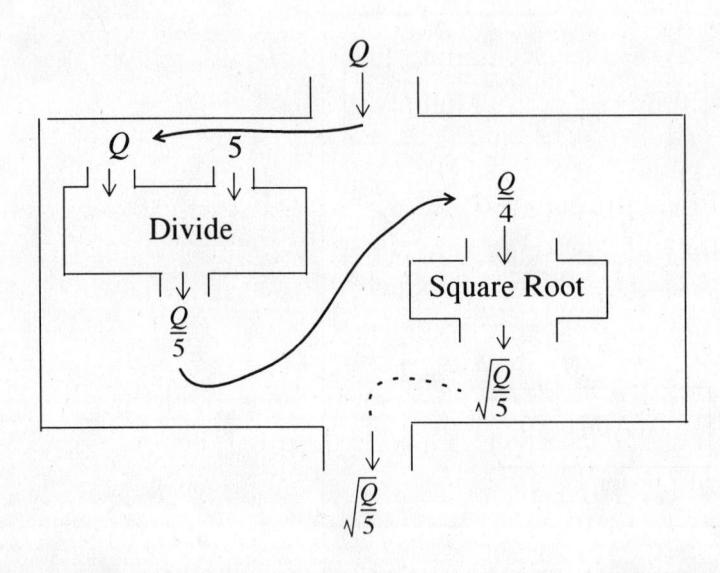

b.

c.

d.

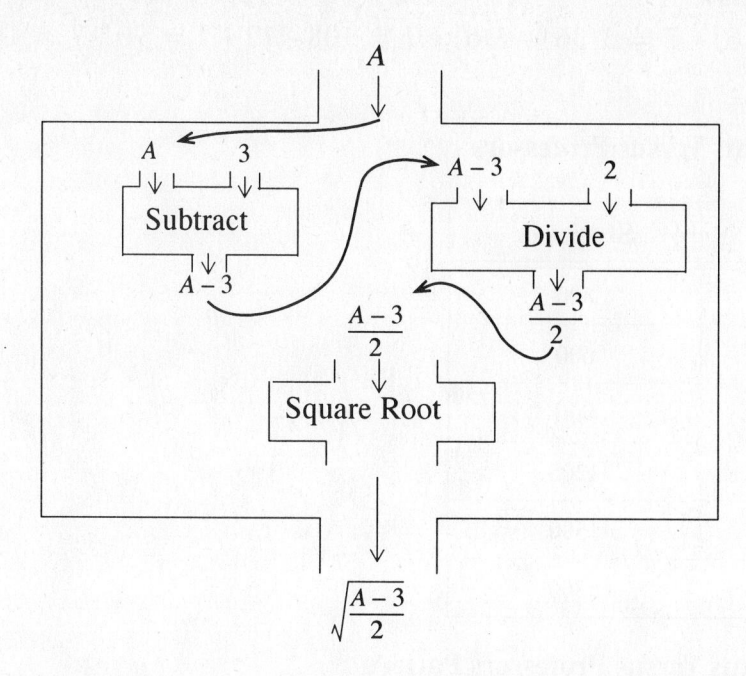

[Diagram: Function machine with input A going into a "Subtract" box (with 3), outputting A − 3, which feeds into a "Divide" box (with 2), outputting (A − 3)/2, which feeds into a "Square Root" box, outputting $\sqrt{\dfrac{A-3}{2}}$.]

4. a. If $N = 13b$ and $b = 6$, then $N = 13(6) = 78$. Evaluating.

 b. If $N = 13b$ and $N = 936$, then $936 = 13b$. Since the function machine requires us to multiply the input by 13 to get the output, we divide the output by 13 to get the input. So, $b = \dfrac{936}{13} = 72$. Solving.

 c. If $C = 4z^3$ and $z = 7$, then $C = 4(7)^3 = 4(343) = 1372$. Evaluating.

 d. If $Q = 5m^2$ and $m = 12$, then $Q = 5(12)^2 = 5(144) = 720$. Evaluating.

 e. If $Q = 5m^2$ and $Q = 245$, then $245 = 5m^2$. Since the function machine requires us to square the input and multiply the square by five, we must divide by 5 and find the square root of the quotient to get the input when given output. So,
 $m = \sqrt{\dfrac{245}{5}} = \sqrt{49} = 7$. Solving.

 f. If $A = 3 + 2b^2$ and $b = 13$, then $A = 3 + 2(13)^2 = 3 + 2(169) = 3 + 338 = 341$. Evaluating.

 g. If $A = 3 + 2b^2$ and $A = 75$, then $75 = 3 + 2b^2$. The function machine process is square, multiply by 2, add 3. To reverse the process, subtract 3, divide by 2, and find the square root. So, $b = \sqrt{\dfrac{75-3}{2}} = \sqrt{\dfrac{72}{2}} = \sqrt{36} = 6$. We are solving an equation.

 h. If $L = 3r^2 - 2r + 7$ and $r = 6$, then

31

$L = 3(6)^2 - 2(6) + 7 = 3(36) - 2(6) + 7 = 108 - 12 + 7 = 96 + 7 = 103$. Evaluating a function.

5. a. **Table 6: Students Versus Professors**

Professors	Students
50	300
100	600
150	900
200	1200
250	1500
300	1800

b. **Table 7: Students Versus Professors Pattern**

Professors	Students
50	6(50)
100	6(100)
150	6(150)
200	6(200)
250	6(250)
300	6(300)
p	6p

c. $s = 6p$

d. The finite differences between the number of students is 300. They become constant in the first finite difference.

e. As the number of professors increases by 50, the number of students increases by 300. This is verified by the finite differences.

f. The domain and range of the mathematical process is all real numbers.

g. The domain of the problem situation is all whole numbers. The range of the problem situation is all whole numbers that are multiples of 6.

6. a. There are 72 students for each mathematics faculty member.

b. If $M = 12$ then $S = 72(12) = 864$. If there are 12 full–time mathematics faculty, there are 864 students enrolled in mathematics courses.

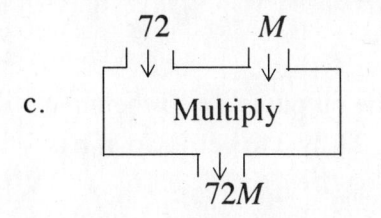

c.
```
     72        M
      ↓         ↓
   ┌─────────────────┐
   │    Multiply     │
   └─────────────────┘
            ↓
           72M
```

d. The domain and range of the mathematical function are both all real numbers.

e. The domain of the problem situation is the set of all whole numbers. The range of the problem situation is the set of all whole number multiples of 72.

7. $\dfrac{S}{E} = \dfrac{720}{18} \Rightarrow \dfrac{S}{E} = 40 \Rightarrow S = 40E$. So $S = 40E$.

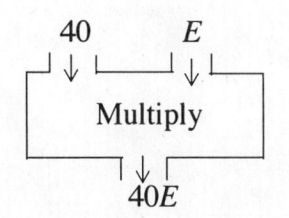

```
     40        E
      ↓         ↓
   ┌─────────────────┐
   │    Multiply     │
   └─────────────────┘
            ↓
           40E
```

8. Let c represent the number of cars parked in the lot and w represent the number of wheels in the lot.

   a. $w = 4c$

   b. The independent variable is c, the number of cars.

   c. The domain and range of the mathematical process is all real numbers.

   d. The domain of the problem situation is the set of whole numbers less than 21. The range of the problem situation is the set of whole numbers less than 81 that are multiples of 4.

9. a. Input: S P P; Output: S P P P P

   b. For example, apply Rule 2 to input S P to get output S P C.

   c. The domain of Rule 1 is the set of all valid paths.

   d. The domain of Rule 2 is the set of all valid paths that end in a P. This is true since Rule 2 cannot be applied if the path does not end in P.

10. Answers will vary.

11. a. Suppose $s = 5t$ defines the function. When the output $s = 35$ the input will be 7.

b. Suppose $k = l + 11$ defines the function. When the output k is 24 the input l will be 23.

c. Suppose $d = 3c - 8$ defines the function. When the output is $d = 13$ the input will be found by adding 8 to 13 to get 21 and then dividing 21 by 3 to get 7. So, if c is 7 then d will be 13.

d. Suppose $y = 7x + 2$. If $y=58$ the input would be $\dfrac{58-2}{7} = 8$.

Section 3.3 Function: Notation for Algebraic Representations

p. 183

2. Note the direction of the arrows.

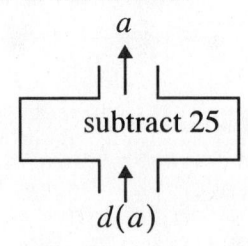

3. a. The unknown is the alarm time a. We must solve the equation $10:00 = a + 25$. To do this we reverse adding 25 by subtracting 25 from the time 10:00. This gives us 9:35. The alarm time should be 9:35 A.M.

b. Here we must determine the value of $d(8:32) = 8:32 + 25 = 9:00$. We will depart at 9:00 A.M.

4. a. The domain of the mathematical process is all real numbers.

b. The domain of the car rental problem is the set of all possible miles that the car could be driven in a given day. This would be the set of whole numbers less than some maximum number of miles.

5. a. The range of the mathematical process is all real numbers.

b. The range of the car rental problem is the set of all possible charges for the car. This would be the set of all numbers starting at 35 in increments of 0.17.

6. a. We solve $11 = r - 4$ by adding 11 and 4 to get $r = 15$.

b. $y(2) = 9(2) = 18$

c. $F(5) = 2(5) - 7 = 3$

d. We solve $14 = 3p + 2$ by subtracting 2 from 14 then dividing the result by 3 to get $p = 4$.

7. a. $R(n) = 19n$

b. $R(2500) = 19(2500) = 47500$. If 2500 tickets are sold, the receipts are $47,500.

c. If the receipts were $59,185 we would have to solve $59185 = 19n$ to determine the number of tickets sold. The solution would be 59,185 divided by 19 or $n = 3115$.

8. a. $s(p) = 6p$

b. $s(75) = 6(75) = 450$. If there are 75 professors, there are 450 students.

c. If there are 486 students there would be 81 students.

9. a. $w(c) = 4c$

b. $w(17) = 4(17) = 68$. If there are 17 cars in the lot, there are 68 wheels in the lot.

10. This is your personal problem.

11. Answers will vary.

12. Some quantities that are constant include $1.36, twenty dollar bill, twenty minutes late, one–half, one–quarter, one–sixteenth, two feet, four feet, $17, six people, three toll booths, one dollar.

13. Some variables include:

amount of gas pumped (input) and gas charge (output);

gas charge (input) and change (output);

number of students enrolled (input) and number of students still remaining (output);

amount of light two feet away from bulb (input) and amount of light one foot away from bulb (output);

how well I could see one foot away (input) and how well I could see four feet away (output);

number of hours I work in the summer (input) and total pay in summer (output);

amount paid at a toll booth (input) and my change (output).

14. amount of gas pumped (input) and gas charge (output): Increasing the input increases the output.

    gas charge (input) and change (output): Increasing the input decreases the output.

    number of students enrolled (input) and number of students still remaining (output): Increasing the input increases the output.

    amount of light two feet away from bulb (input) and amount of light one foot away from bulb (output): Increasing the input increases the output.

    how well I could see one foot away (input) and how well I could see four feet away (output): Increasing the input increases the output.

    number of hours I work in the summer (input) and total pay in summer (output): Increasing the input increases the output.

    amount paid at a toll booth (input) and my change (output): Increasing the input decreases the output.

15. a. 4^{26} b. 4^{153}

    c. not possible   d. x^{15}

    e. 1   f. not possible

    g. x^8 h. not possible

    i. not possible   j. 11^2

16. a. $7x + 10$ b. $5x^2 + 4x + 9$

    c. $15y^2 - 24y$ d. $16x^2 + 24x + 9$

17. a. $3(3b + 2)$ b. prime

    c. $4y(3y^2 + y + 7)$

Section 3.4 Function: Geometric Interpretations

p. 197

2. Site of Worship: (4, 7), (4, 8), (4, 9), (4, 10).

   Site of Ancient Secrets: (7, 6), (8, 6), (9, 6).

Math Site 1: (5, 1), (6, 1). Math Site 2: (1, 3), (2, 2).

3. Let the points be labelled: $A(30, 2)$; $B(50, 7)$; $C(10, 9)$; $D(60, 0)$; $E(35, 6)$; $F(0, 4)$.

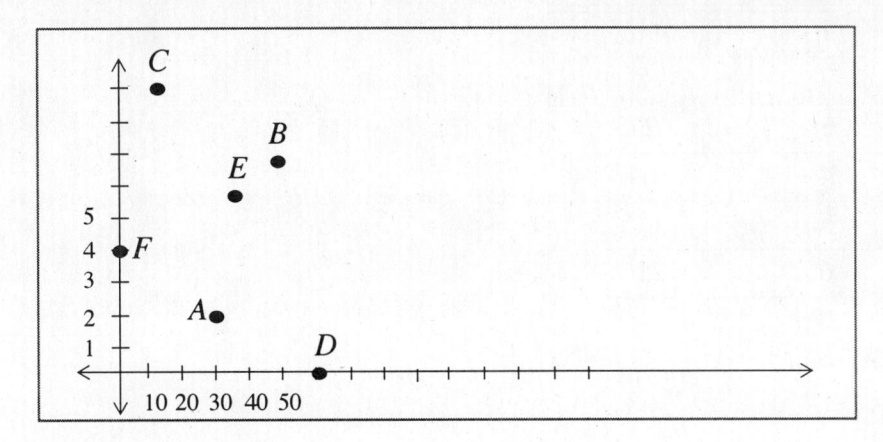

4. a. **Professors and Students**

Professors	Students
10	60
20	120
30	180
40	240
50	300
60	360

b. $S(P) = 6P$

c.

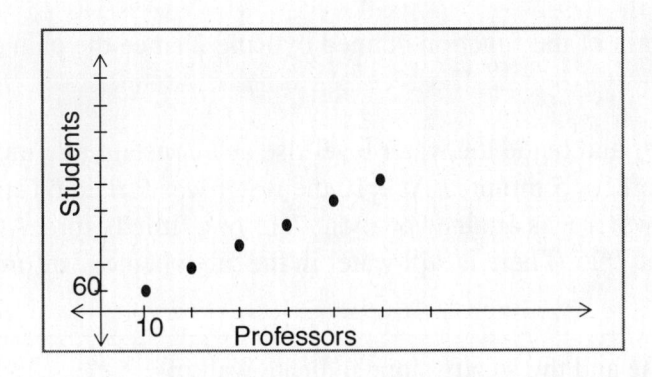

5. a. **Cubes**

Number	Cube
1	1
2	8
3	27
4	64
5	125
6	216

b. $C(n) = n^3$

c.

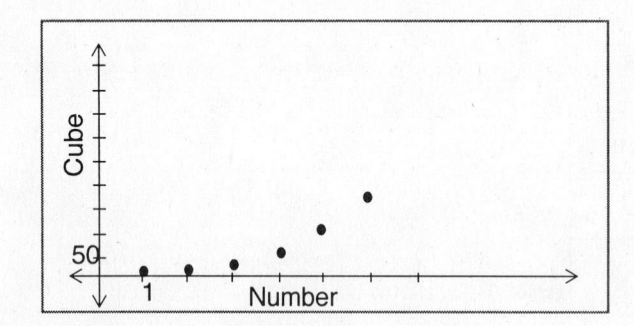

6. a. **Table 3: Rule 2 of SPC**

Input	Output
S P P	S P P C
S C P	S C P C
S P C C P	S P C C P C

b. S P C is not in the domain of the function defined by Rule 2 since the path does not end in P.

7. Someone turned on the water and made the water level rise at a constant rate until 7:05 when the water was turned off for 5 minutes. At 7:10 the water was turned on stronger until the desired amount of water was attained at about 7:15 or a little before. At 7:20 the drain was opened until about 7:25. There is still water in the tub when our information ends at 7:30.

8. a. *B*: the relatively long time and the steady slope indicate walking.

   b. *A*: the quicker pace and varying speeds suggest a leisurely bike ride.

c. *C*: the rapid change in distance during certain time intervals suggests running.

d. The students stopped for a rest.

9. a. Only Graph *C* could represent a journey.

b. Graph *A* indicates that something moved from the starting point to some positive distance from the starting point instantaneously then stayed there.

Graph *B* may be two graphs which describe two points coming together at a certain time and distance from the beginning. One starting at the starting point the other some distance away from the beginning.

Graph *C* shows the path of someone who starts at the starting point and goes a certain distance away, at the first corner starts going towards the starting point, at the second corner resumes proceeding away from the starting point at a greater speed than on the first or second leg.

10. {97}

11. The set of prime numbers less than 100 is not closed under addition since you can add a pair of prime numbers like 3 and 5 and get an answer, 8, that is not prime.

12. A variable can be used to represent a quantity that does not change from one time to the next. The variable x represents the string C in the path S C P under Rule 2.

A variable can be used to generalize such as $a + 0 = a$.

A variable can be used to represent any value in a given replacement set. For example, the variable n in the function $R = 19n$ represented the number of concert tickets sold. This could be any whole number less than or equal to some number that represents the maximum seating capacity.

13. a. **Yards to feet**

Yards	Feet
1	3
2	6
3	9
4	12
5	15

b. $F(Y) = 3Y$.

14. $2x - 5 = 17$ for example.

15. a. $x = 7$ b. $y = 12$
    c. $t = 5$ d. $x = 0$

16. a. $(0, 2)$
    b. $(5, 0), (8, 0)$

Section 3.5 Triangular Numbers

p. 216

2. If $n = 7$ then $T = \frac{1}{2}(7)(7+1) = 28$. Yes, the answer agrees with the table value.

3. Yes. The generalization does not depend on the number of numbers being even or odd.

4. If $n = 113$ then $T = \frac{1}{2}(113)(113+1) = 6441$.

5. If $n = 2173$ then $T = \frac{1}{2}(2173)(2173+1) = 2,362,051$.

6. The nth triangular number equals one-half the product of n and one more than n.

7. We must find the value of n that makes $\frac{1}{2}(n)(n+1) = 465$. If we use the table display on our calculator to examine input-output pairs for the triangular number function we can scroll until we find 465 in the output column. The corresponding input value is 30. In other words we have found the solution to our equation using our calculator's table feature.

8. a. Table 4: Numeric Representation of Triangular Numbers

Finite Differences of Index	Index	Triangular Number	Finite Differences in Triangular Number	Second Finite Differences in Triangular Number
2 – 1 = 1	1	1	3 – 1 = 2	3 – 2 = 1
3 – 2 = 1	2	3	6 – 3 = 3	4 – 3 = 1
1	3	6	4	1
1	4	10	5	1
1	5	15	6	1
1	6	21	7	1
1	7	28	8	1
1	8	36	9	1
1	9	45	10	
	10	55		

b. In the concert receipts problem the first finite differences are all the same.

c. $T(n) = \dfrac{n^2 + n}{2}$ The variable appears to the second power in this function.

d. The functions like $R(n) = 19n$ are defined by multiplying the variable by a constant. The variable appears only raised to the first power. This makes the first finite differences constant. The graphs of such functions follow a straight line. This is an example of a linear function.

Functions like $T(n) = \dfrac{n^2 + n}{2}$ or $S = n^2$ contain the variable raised to the second power. The first finite differences computed in a table are not constant but the second finite differences are. The graph of such functions follow curves called parabolas. These are examples of quadratic functions.

9. a. 1, 4, 9, and 16.

b. 25.

c. 36, 49, 64, 81, 100. Hopefully you recognize the pattern of squares.

d. **The Squares**

Dots on a side	Total number of dots
1	1
2	4
3	9
4	16
5	25
6	36
7	49
8	64
9	81
10	100

e. If $n = 4$ then $S = 16$. If $n = 8$ then $S = 64$

f. $S = n^2$

g. The graphic display produced on a TI-83 together with the viewing window is shown. The table columns are stored in list number 1 and 2 in the calculator. One of the points is displayed using the trace feature of the calculator.

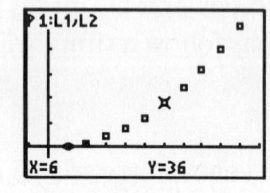

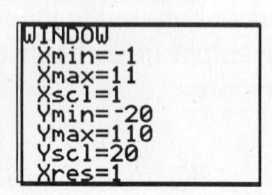

h. The graph is more like the graph of the triangular numbers function. They both follow the shape of a curve rather than a straight line as in the graph of the concert receipts function.

Section 3.6 Power and Factorial Functions

p. 230

2. a. The finite differences become constant after two computations.

b. The model is based on squares since the second finite difference is constant.

c. $D(t) = 16t^2$.

3. a. The labels are a help. It looks like $S(x)$ does not appear in Figure 7. There does seem to be a curve that the three graphs follow. $f(x)$ curves up the steepest.

b. $f(x)$ increases most rapidly followed by $F(x)$ then by $G(x)$ the next steepest would be $S(x)$ if it were shown.

c. They would appear to be straight lines because of the scale used on the vertical axis. Neither grow above 1000 when the input is 10.

d. We would only see the first three points. The other points would be off the top of the screen.

e. The point (1, 1) is on the graph of all three functions.

4. There are at most two students in the room. Can you figure out why?

5. a. 9! = 362,880. There are 362,880 different arrangements of the players in the nine positions.

b. 362,880 innings are necessary to try all possible arrangements.

c. $\dfrac{362,880}{6} = 60,480$. 60,480 six–inning games are required to try all possible arrangements.

d. $\dfrac{60,480}{365} \approx 165.7$ Approximately 165.7 years are required to try all arrangements.

Section 3.7 Review

p. 233

1. a. $N = 17(4) = 68$; Evaluating.

b. b is a little bigger than 43.4. To find b we divide 738 by 17. Solving.

c. $Q = 60.75$; Evaluating.

d. $m = 9$ or $m = -9$; Solving.

e. $A = 725$; Evaluating.

f. $b = 7$; Solving.

g. $L = 72$; Evaluating.

2. a. b, m, and r are the independent variables.

   b. $N, Q, A,$ and L are the dependent variables.

3. a. For N multiply by 17.   c. For Q first square then multiply by 3.

   e. For A square, multiply by 2 then add 3.

4. a. 33

   b. First cube, multiply by 4, divide the 8 by 4, then add and finally subtract.

   c. $3 + (4(2^3) - 8 \div 4)$

5. a. m the number of miles driven.   b. $C(m)$ the cost.

   c. $C(m) = 0.17m + 27$

   d. $C(137) = 0.17(137) + 27 = 50.29$ The cost is $50.29. We evaluated a function to find this cost.

   e. $m = \dfrac{42.81 - 27}{0.17} = 93$ We solve the equation $0.17m + 27 = 41.81$ to find that we drove 93 miles.

6. a. **Table 1: Foreign Versus U.S.-Made Cars**

Foreign-Made Cars	U.S.-Made Cars
3	12
5	20
7	28
9	36

   b. The input would be the number of foreign-made cars and the output is the number of U.S.-made cars.

   c. $A = 4F$

d. The following graph was produced on a TI-83 graphing calculator.

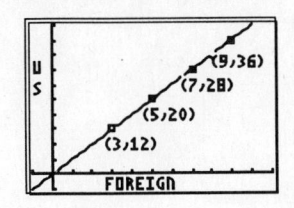

7. a.

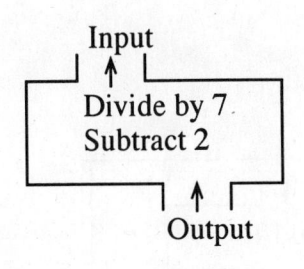

    Input
    ↓
    Divide by 7
    Subtract 2
    ↓
    Output

    b. $7(13) + 2 = 93$

    c. $x = 1$

8. a.

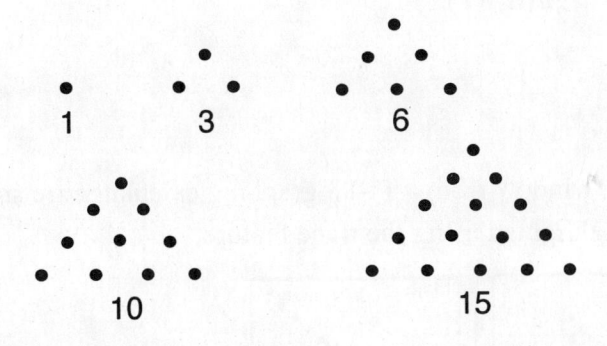

    1   3   6

    10   15

    b. **The Triangular Numbers**

n	T(n)
4	10
5	15
6	21
7	28

n	T(n)
8	36
9	45
10	55
11	66
12	78

c.

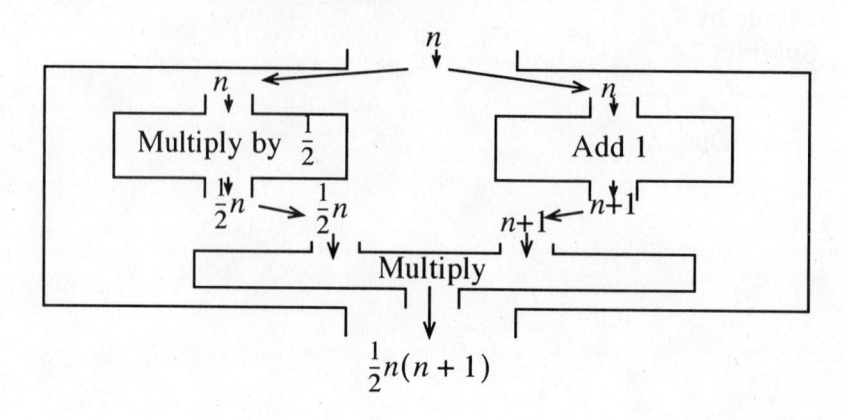

$$n \to \text{Multiply by } \tfrac{1}{2} \to \tfrac{1}{2}n$$
$$n \to \text{Add } 1 \to n+1$$
$$\tfrac{1}{2}n, \, n+1 \to \text{Multiply} \to \tfrac{1}{2}n(n+1)$$

d. $T(n) = \dfrac{1}{2}n(n+1)$

e. The graph and the viewing window from a TI-83 graphing calculator are shown. The point (10, 55) has been highlighted using the trace feature.

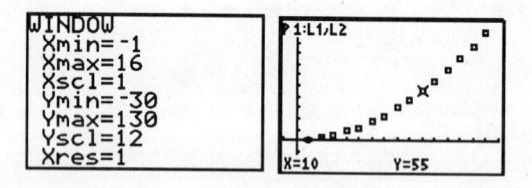

```
WINDOW
Xmin=-1
Xmax=16
Xscl=1
Ymin=-30
Ymax=130
Yscl=12
Xres=1
```

1:L1,L2

X=10   Y=55

9. a. 28 appears in the table.

b. It might be helpful to refer to the dot pictures of the first five triangular numbers to explain this.

c. T increases by a greater and greater amount. This can be seen from the graph since the points plotted go up as the first coordinate increases. It is also evident by looking at the table. One might include the first finite differences in another column to show that the outputs increase by a greater and greater amount.

d. $T(473) = \dfrac{1}{2}(473)(473 + 1) = 112101$. The algebraic representation using the order of operations indicated in the function machine or a calculator to make the computation seems appropriate here.

e. 153 is the 17th triangular number. Extending the table to an input of 17 reveals this. The display of such a table is shown below together with the Y= menu and the TABLE SETUP screen used to produce the table.

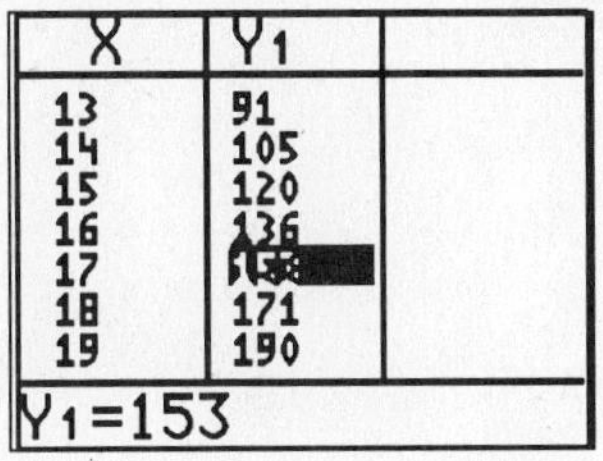

10. It seems to depend on the problem question, doesn't it?

11. a. $2p = 36$
$p = 18$

b. $p^2 = 36$
$p = 6$ or $p = -6$

c. $p + 11 = 36$
$p = 25$

d. $4p - 7 = 36$
$4p = 29$ Not a whole number.
$p = 4\frac{1}{7}$

e. $3 + 2p = 36$
$2p = 33$
$p = 16\frac{1}{2}$

Not a whole number.

f. $p^2 + 3 = 36$
$p = 33$
$p = \sqrt{33}$ or $p = -\sqrt{33}$

Not a whole number.

12. a.

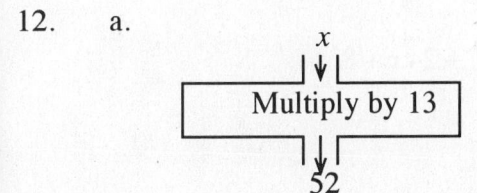

x
↓
[Multiply by 13]
↓
52

$x = 4$

b.

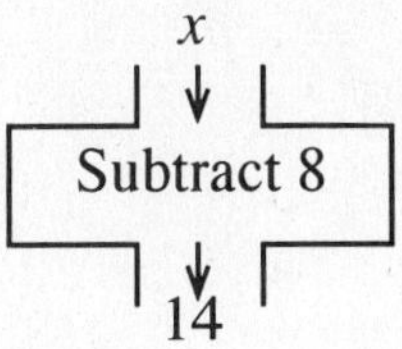

x
↓
[Subtract 8]
↓
14

$x = 22$

c.

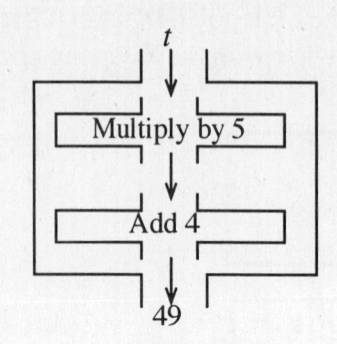

$5t + 4 = 49$
$5t = 45$
$t = 9$

[flowchart: t → Multiply by 5 → Add 4 → 49]

d.

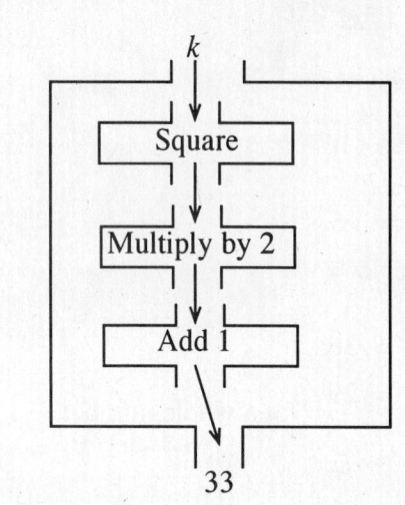

$2k^2 + 1 = 33$
$2k^2 = 32$
$k^2 = 16$
$k = 4$ or $k = -4$

[flowchart: k → Square → Multiply by 2 → Add 1 → 33]

13. a. 3^{12} b. $(5 \cdot 7)^3$

    c. 1   d. 1

    e. x^5 f. z^{99}

14. a. $3x^2 + 7x + 6$ b. $15x^3 + 16x^2 + 24x + 8$

    c. $16y^2 + 56y + 49$ d. $5x + x^2$

15. a. $2 \cdot 3 \cdot 5^2$ b. $3^2 \cdot 7 \cdot 11^3$

    c. $11 \cdot 13 \cdot 17$ d. $2^6 \cdot 13$

    e. $2^4 \cdot 3^2 \cdot 5 \cdot 7$ f. $2^7 \cdot 3^2 \cdot 5 \cdot 7$

16. a. prime   b. $2(a^2 + 2a + 3)$

    c. $x(5 + x)$

CHAPTER 4

Integers: Expanding on a Mathematical System

Section 4.1 Integers and the Algebraic Extension

p. 255

2.   a.   –6   b.   5

     c.   0   d. $-5x$

     e. $3y$ f. $-(x+y)$ or $-x-y$

3.   a.   The whole numbers are closed under the operations of addition and multiplication.

     b.   The integers are closed under the operations of addition, subtraction, and multiplication.

4.   a. $-t = -7$ b. $-b = -(-2) = 2$

     c. $-r = -823$

5.   a.   In $-12-5$ the first "–" is used to represent negative 12. The second represents subtraction with inputs -12 and 5.

     b.   In $-12-(-5)$ the first minus sign represents negative 12 and the second represents subtraction with input -12 and -5. The final minus sign is used to indicate negative 5.

     c.   In $12-5$ the minus sign represents subtraction with inputs 12 and 5.

     d.   The two minus signs negative 12 and negative 5.

     e.   The minus in front of the x represents taking the opposite of the value of x.

     f.   This one means subtraction with inputs x and y.

     g.   The two minus signs represent taking the opposite of x and y.

     h.   The first and the last minus signs represent taking the opposite of x and y. The other one indicates subtraction with inputs $-x$ and $-y$.

     i.   The first represents negative 5, the last indicates taking the opposite of x and the one in the middle means subtract $-x$ from -5.

j. The second minus sign is used to represent negative 5. The first represents subtraction with inputs x and -5.

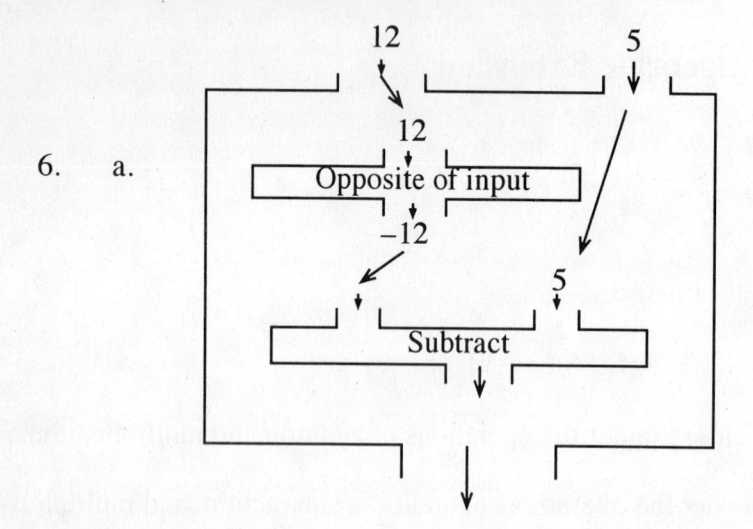

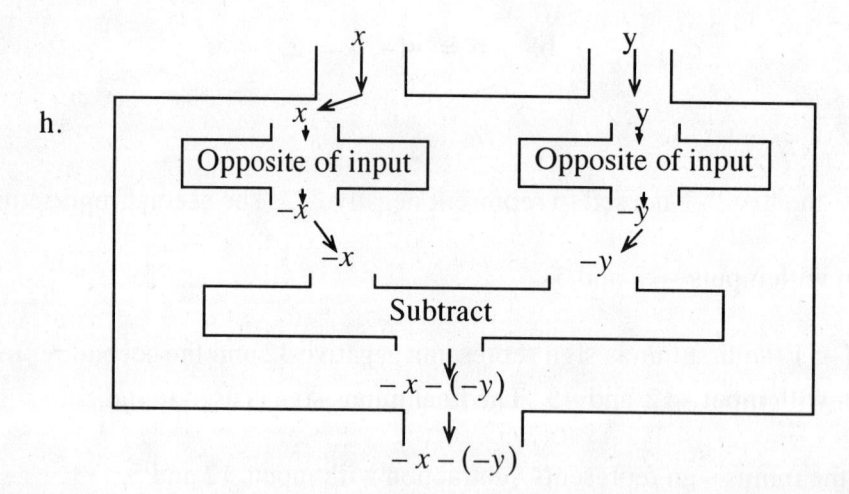

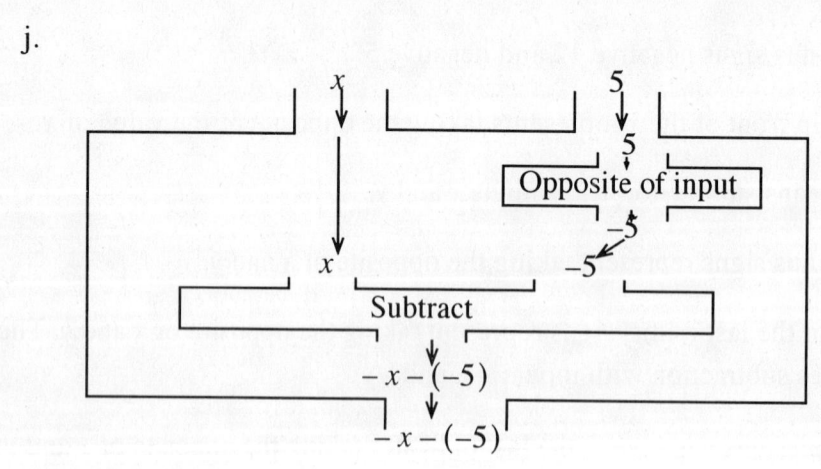

6. a.

    h.

    j.

50

7. a. $int(rand \cdot 10) + 1$

   b. *rand* is between 0 and 1 exclusive. Seven times *rand* is between 0 and 7 exclusive. $7rand - 4$ is between −4 and 3 exclusive. $int(7rand - 4)$ is between −4 and 2 inclusive. $int(7rand - 4) + 1$ is between −3 and 3 inclusive.

8. a.

Pounds of steak	Pounds of hamburger
10	60
20	120
30	180
40	240
50	300
60	360

   Table 1: Steak and hamburger

   b. $H = 6S$ where S represents the number of pounds of steak and H represents the number of pounds of hamburger.

   c.

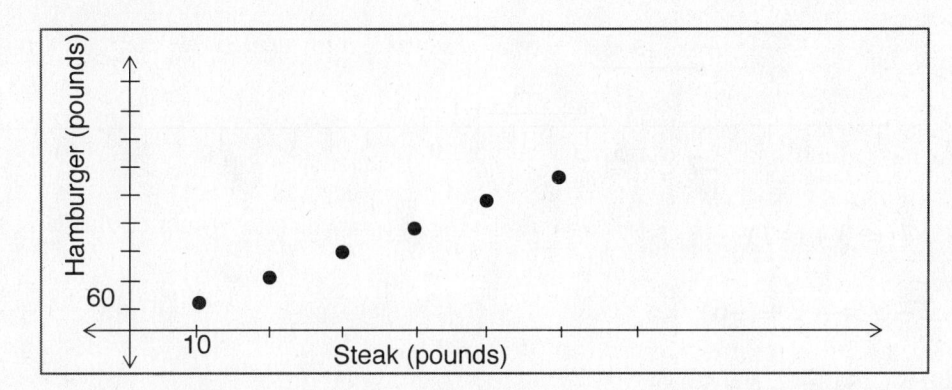

Section 4.2 Operations on Integers

p. 275

2. a.

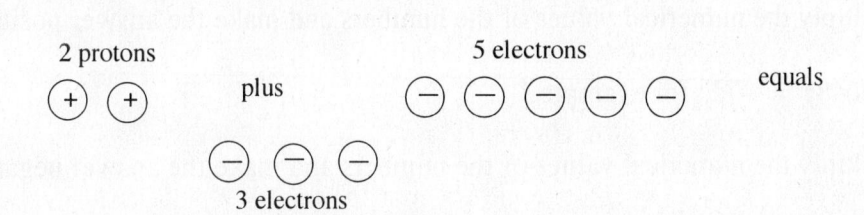

51

b. 7 electrons plus 3 electrons equals 10 electrons

c. 9 protons plus 2 electrons equals 7 protons

3. a.

    b.

    c.

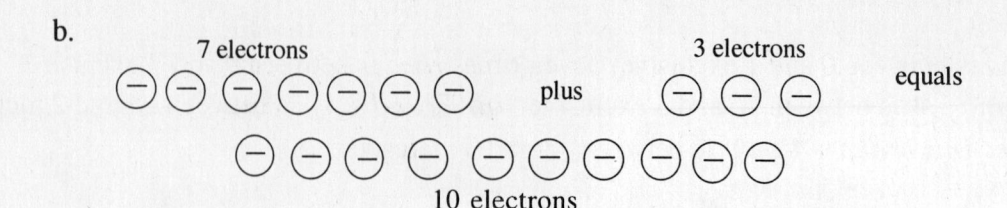

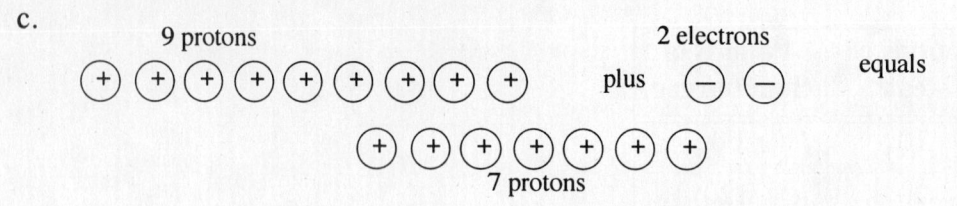

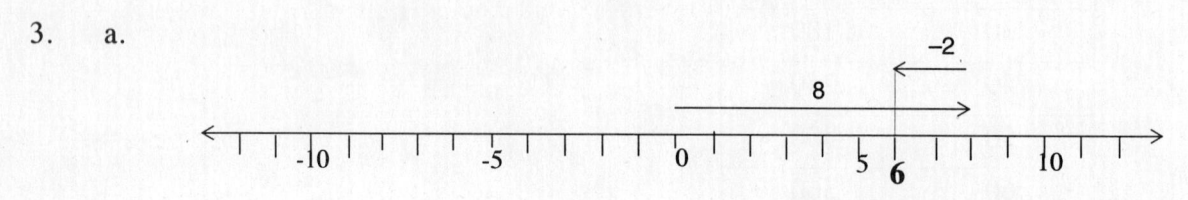

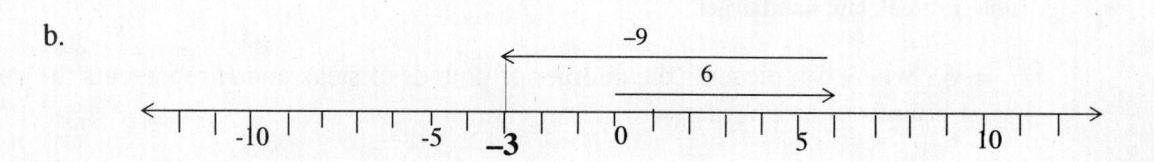

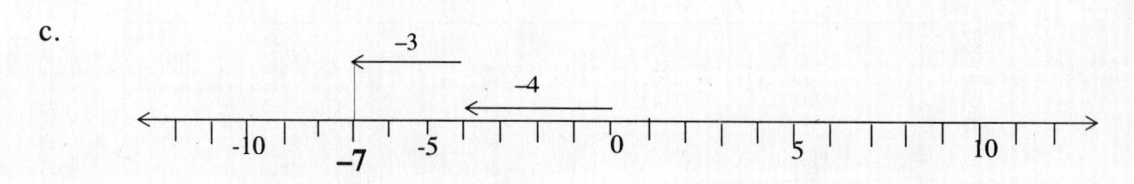

4. a. $3 - 7 = 3 + (-7)$

    b. $-2 - 6 = -2 + (-6)$

    c. $-4 - (-5) = -4 + 5$

    d. $-9 - (-7) = -9 + 7$

5. a. $(-9)(-8) = 72$, for example.

    b. Multiply the numerical values of the numbers and make the answer positive.

6. a. $(9)(-8) = -72$, for example.

    b. Multiply the numerical values of the numbers and make the answer negative.

7. a. $(-18) \div (-3) = 6$, for example.

b. Divide the numerical values of the numbers and make the answer positive.

8. a. $(18) \div (-3) = -6$ or $(-18) \div 3 = -6$, for example.

   b. Divide the numerical values of the numbers and make the answer negative.

9. a. $-8 - 7 = -8 + (-7) = -15$.

   b. $(-8)(-7) = 56$.

   c. $-(-35) = 35$.

   d. $-15 \div 3 = -5$

   e. $-13 + 2 = -11$

   f. $6 + (-9) - 5 - (-7) = 6 + (-9) + (-5) + 7 = -3 + (-5) + 7 = -8 + 7 = -1$.

   g. $(-3)(8) + (-2)(-6) = -24 + 12 = -12$

   h. $(8 + (-2))(5 - 9) = (6)(-4) = -24$.

   i. $17 - (-9) + (-54) - 38 = 17 + 9 + (-54) + (-38) = 26 + (-54) + (-38) =$
      $-28 + (-38) = -66$.

   j. $-24 \div (-6)(-3) = 4(-3) = -12$.

   k. $4(-3)^2 - 2(2)^3 + (-3)(5)) = 4(9) - 2(8) + (-15) = 36 - 16 + (-15) =$
      $20 + (-15) = 5$.

10. a. Both answers are 3.

    b. Both answers are −4.

    c. Both answers are −10.

    d. The commutative property of addition holds for integers.

11. a. The answers are 2 and −2.

    b. The answers are 8 and −8.

    c. The answers are 3 and −3.

    d. Subtraction of integers is not commutative.

12. a. Both answers are –6.

 b. Both answers are –19.

 c. Both answers are 12.

 d. The associative property of addition holds for integers.

13. a. The answers are –5 and 7.

 b. The answers are –17 and –1.

 c. The answers are 22 and 12.

 d. Subtraction of integers is not associative.

14. a. Case 1: Bathtub filling, play videotape forward.

 Case 2: Bathtub filling, play videotape backward.

 Case 3: Bathtub emptying, play videotape forward.

 Case 4: Bathtub emptying, play videotape backward.

 b. **Table 4 Bathtub Model**

Bathtub filling or emptying	Videotape direction	What you see on the screen
+	+	+
+	–	–
–	+	–
–	–	+

 c. The product of two numbers whose signs are both the same is a positive number and the product of two numbers whose signs are different is a negative number.

15. a. In – 12 – 7 the first "–" is used to represent negative 12. The second represents subtraction with inputs –12 and 7.

 b. In – 12 – (–7) the first minus sign represents negative 12 and the second represents subtraction with input –12 and –7. The final minus sign is used to indicate negative 7.

 c. In 12 – 7 the minus sign represents subtraction with inputs 12 and 7.

d. The two minus signs are used to represent negative 12 and negative 7.

e. The minus in front of the *3x* represents taking the opposite of the value of *3x*.

f. This one means subtraction with inputs *3x* and *y*.

g. The two minus signs represent taking the opposite of *3x* and *y*.

h. The first and the last minus signs represent taking the opposite of *3x* and *y*. The other one indicates subtraction with inputs $-3x$ and $-y$.

i. The first represents negative 9, the last indicates taking the opposite of *x* and the one in the middle means subtract $-x$ from -9.

j. The second minus sign is used to represent negative 9. The first represents subtraction with inputs *x* and -9.

16. a. $-x^2 + 3x - 2$

    b. $(3x - 8) - (2x + 3) = 3x - 8 - 2x - 3 = x - 11$

    c. $(x^2 + 9x - 3) - (4x^2 + 3x + 2) = x^2 + 9x - 3 - 4x^2 - 3x - 2 = -3x^2 + 6x - 5$

    d. $6x^3 - 15x^2$

    e. $-10x^3 + 35x^2 - 45x$ f. $x^2 - 5x - 36$

    g. $12x^2 + 13x - 14$ h. $x^2 - 49$

    i. $16x^2 - 9$ j. $4x^2 - 44x + 121$

    k. $2x^4 - 11x^3 - 29x^2 - 2x + 5$ l. $x^4 - 1$

17. a. $T(-3) = 5(-3) - 2 = -15 - 2 = -17$

    b. $5r - 2 = -17$
    $5r = -15$
    $r = -3$

18. If $y(x) = 3x^2 - x - 4$ then $y(-5) = 3(-5)^2 - (-5) - 4 = 3(25) + 5 - 4 =$

    $75 + 5 - 4 = 80 - 4 = 76$

19. a. $\begin{aligned} 5x - 2 &= -12 \\ 5x &= -10 \\ x &= -2 \end{aligned}$
    b. $\begin{aligned} 3a + 7 &= -17 \\ 3a &= -24 \\ a &= -8 \end{aligned}$

   c. $\begin{aligned} 9q + 1 &= 46 \\ 9q &= 45 \\ q &= 5 \end{aligned}$
    d. $\begin{aligned} -2x - 9 &= -21 \\ -2x &= -12 \\ x &= 6 \end{aligned}$

   e. $\begin{aligned} -4c + 5 &= 13 \\ -4c &= 8 \\ c &= -2 \end{aligned}$
    f. $\begin{aligned} -5w + 14 &= 14 \\ -5w &= 0 \\ w &= 0 \end{aligned}$

   g. $\begin{aligned} 5 - 8x &= -27 \\ -8x &= -32 \\ x &= 4 \end{aligned}$
    h. $\begin{aligned} -7 + 3y &= -28 \\ 3y &= -21 \\ y &= -7 \end{aligned}$

   i. $\begin{aligned} -1 - 6x &= 23 \\ -6x &= 24 \\ x &= -4 \end{aligned}$
    j. $\begin{aligned} -4 - 5p &= -69 \\ -5p &= -65 \\ p &= 13 \end{aligned}$

Section 4.3 The Absolute Value Function

p. 295

2. a. x_1
   b. x_2
   c. $\Delta x = x_2 - x_1$

3. a. A negative change would imply a decrease in petty cash.
   b. A positive change would imply an increase in petty cash.

4. **Table 7: Daily Change in Petty Cash Account**

Initial Value x_1 in $	Final Value x_2 in $	Change in Petty Cash Δx in $
5	3	−2
3	5	2
3	3	0
0	7	7
7	0	−7
0	−2	−2
−2	0	2
−3	4	7
4	−3	−7
1	−1	−2
−6	−3	3
t	t	0
t	t+4	4
t+4	t	−4
−5	−5	0
anything	anything	0
x	x+5	5
x+5	x	−5

5. a. $\Delta x = 0$ No sign.   b. The change would be positive.

    c. The change would be negative.

6. a. $\Delta t = -12$ At 4:00 P.M. the temperature is 15°.

    b. $\Delta t = -9$ At 9:00 A.M. the temperature was 36°, assuming the temperature dropped 3° each hour from 9:00 A.M. until noon.

7. At the beginning of the week, I had twenty–seven minutes of leisure time per day. However, I became a member of a committee on Tuesday which will leave me with only 12

minutes of leisure time per day. Let the variable l_1 represent the amount of daily leisure time at the beginning of the week and l_2 be the amount of daily leisure time after Tuesday. Then Δl represents the change in my leisure time where $\Delta l = l_2 - l_1 = 12 - 27 = -15$.

My leisure time per day has decreased by fifteen minutes.

8. The absolute value of a number indicates the magnitude (size) of the number. When we create a number line, we start at the point labelled zero. We then mark off units to the right and to the left of zero. These units represent how far the point is from zero. If we ignore direction, the units represent a magnitude and that is exactly what absolute value represents. Distance is usually used to represent "how far" without considering direction. "How far" is a non–negative idea, while direction is often indicated by a positive or negative sign.

9. $\Delta m = 109 - 143 = -34$. He had driven $|\Delta m| = |-34| = 34$ miles.

10. a. $|6|$ is the distance between 6 and 0 on the number line. $|-43|$ is the distance between -43 and 0 on the number line.

    b. The expression is $|8|$.

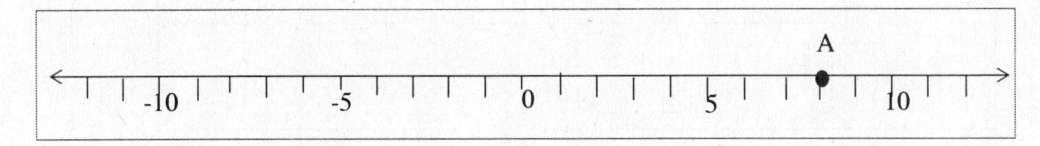

    c. The expression is $|-2|$.

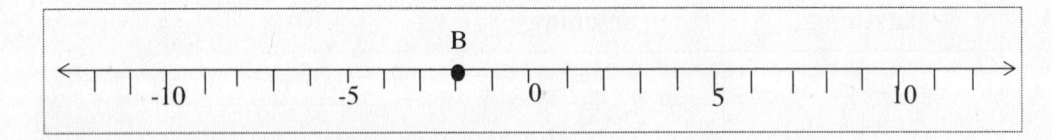

11. a.

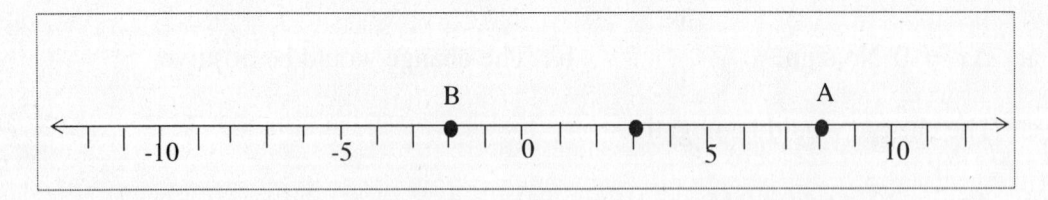

    b. For point A, $|8 - 3| = 5$ For point B, $|-2 - 3| = 5$.

12. a.

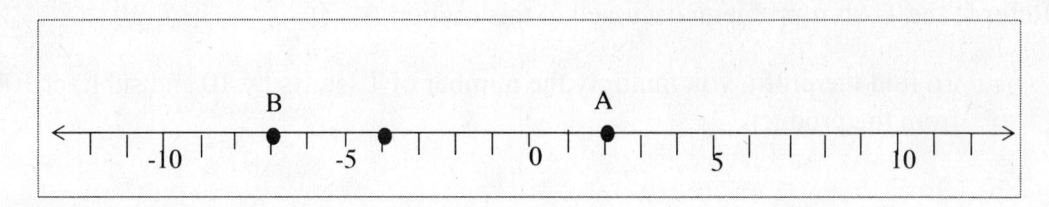

   (number line with B at -7 and A at 2, between -10 and 10)

   b. For point A, $|2 - -4| = 6$ For point B, $|-7 - -4| = 3$.

13. a. -1 and 2 both produce an output of 3.

   b. -3 and 4 both produce an output of 7.

   c. No number produces and output of -1.

14. a. 12 and -2 are the solutions.   b. -4 is the only solution.

   c. The equation has no solutions.   d. -8 and 14 are the solutions.

Section 4.4 Graphing with Integers

p. 306

2. Site of Worship: $(-1, 2), (-1, 3), (-1, 4), (-1, 5)$.

   Site of Ancient Secrets: $(2, 1), (3, 1), (4, 1)$.

   Math Site 2: $(0, -4), (1, -4)$.

   Math Site 1: $(-4, -2), (-3, -3)$.

3. Label the ordered pairs as follows: A(–5, 40), B(3, –50), C(–2, –10), D(6, 20), E(0, –30), F(–4, 0), G(7, 0), and H(0, 10).

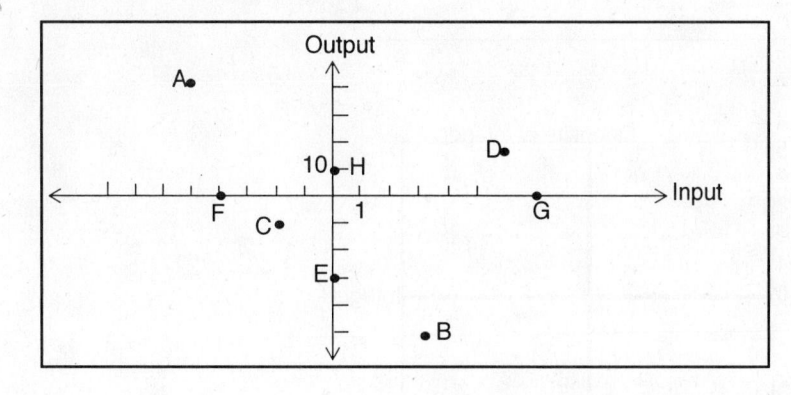

59

4. Refer to the T-shirt problem discussed in this section.

   a. To find the profit, you multiply the number of T-shirts by 10 and subtract 1000 from the product.

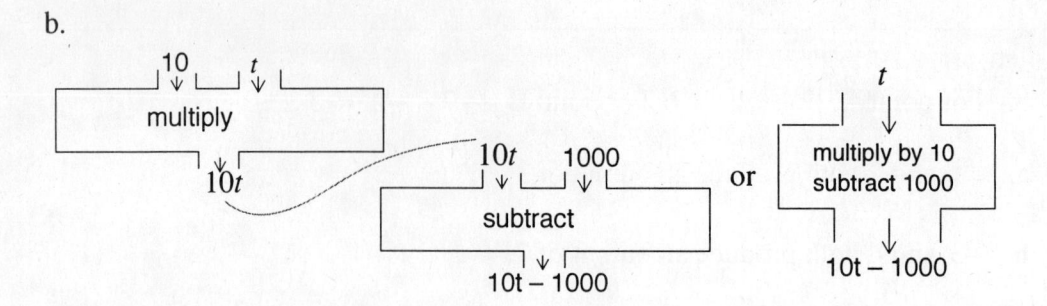

   b.

   c. Let t represent the number of T-shirts sold and P represent the profit. Then $P(t) = 10t - 1000$.

7. **Table 3 Opposites of Integers**

Integer	Opposite of integer
−5	5
−4	4
−3	3
−2	2
−1	1
0	0
1	−1
2	−2
3	−3
4	−4
5	−5

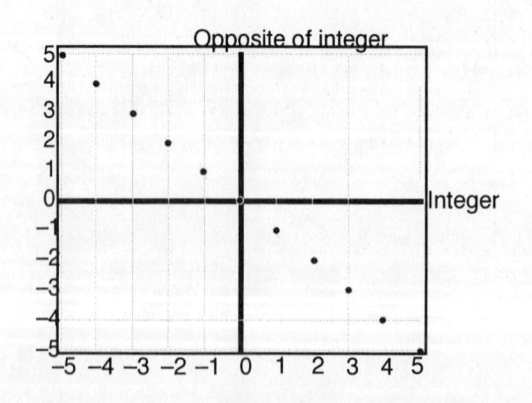

8. **Table 4 Absolute Value of Integers**

Integer	Absolute value of integer
–5	5
–4	4
–3	3
–2	2
–1	1
0	0
1	1
2	2
3	3
4	4
5	5

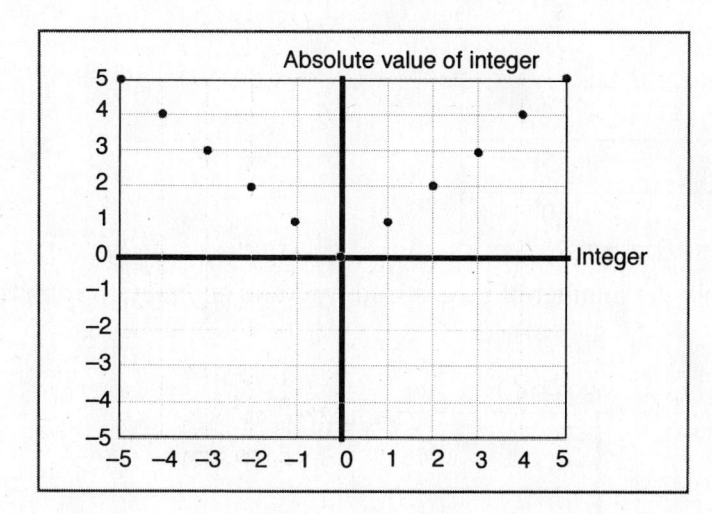

9. a. $-5 + 17 = 12$ b. $2 - 11 = -9$

 c. $13 + (-8) = 5$ d. $-5 - 9 = -14$

 e. $-7 - (-3) = -4$ f. $|-9| = 9$

 g. $-|5| = -5$ h. $-|-6| = -6$

 i. $(-7)(4) = -28$ j. $(-8)(-3) = 24$

 k. $\dfrac{18}{-9} = -2$ l. $-\dfrac{18}{9} = -2$

m. $\dfrac{-18}{9} = -2$ n. $\dfrac{-18}{-9} = 2$

o. $-\dfrac{-18}{-9} = -2$ p. $|-3| - 5 = -2$

q. $-4(|-8|) = -32$ r. $-|-2| + |-9| = 7$

10. The function $P(m) = 3m - 50$ expresses the profit P (in dollars) from the sale of costume jewelry as a function of the number m of pieces sold.

   a. **Profit on sale of costume jewelry**

Pieces sold	Profit ($)
5	−35
10	−20
15	−5
20	10
25	25
30	40

   b. Multiply the number of pieces sold by 3 and subtract 50 from the product.

   c.

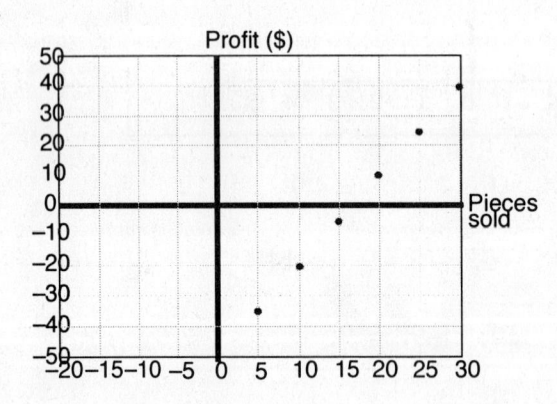

   d. The profit will be a little less than $104 when 51 pieces of jewelry are sold.

11. a. The function is defined by $x - 2$ and the output is -7. $x = -5$

   b. The function is defined by $5y - 3$ and the output is -18. $y = -3$

c. The function is defined by $4 - 3z$ and the output is -17. $z = 7$

d. The function is defined by $-7x - 1$ and the output is 62. $x = -9$

e. The function is defined by $|x - 5|$ and the output is 23. $x = -18$ or $x = 28$

f. The function is defined by $|2x - 1|$ and the output is 5. $x = 3$ or $x = -2$

g. The function is defined by $|4x + 3|$ and the output is -3. There is no input which would produce this output.

h. The function is defined by $|x + 7|$ and the output is 0. $x = -7$

Section 4.5 Using a Graphing Utility

p. 321

2. Xmin = -4, Xmax = 5, Xscl = 1, Ymin = -200, Ymax = 140, Yscl = 20.

3. a. $x = 7.0212766$ and $y = 5.1612903$ on the TI–82 or TI–83.

   b. $(0, 0)$ on the TI–82 and TI-83.

4. The x values are now multiples of five rather than rational numbers. All the values in Quadrants I and IV obtained by TRACE are reasonable values for the problem situation.

5. a. If $P(x) = 10x - 1000$ then $P(132) = 10(132) - 1000 = 320$. This suggests that selling 132 T–shirts results in a profit of $320. The profit indicated in Figure 5 was about $324.

   b. The parametric graph appears below

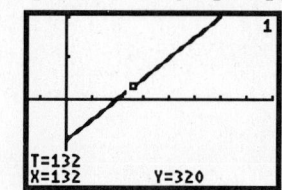

   T=132
   X=132   Y=320

   The result matches the answer to a.

63

6. a. **Professors and Students**

Professors	Students
100	600
200	1200
300	1800
400	2400
500	3000
600	3600

c.

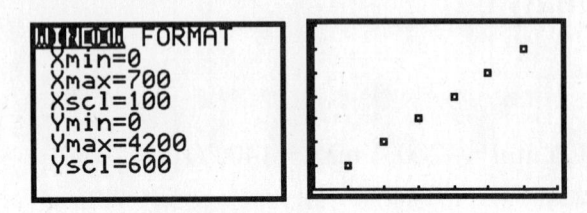

d. The algebraic representation is $S(P) = 6P$.

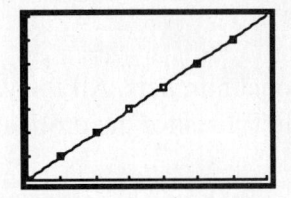

e. **More Professors and Students**

Professors	Students
350	2100
357.44681	2144.6809
364.89362	2189.3617
372.34043	2234.0426

f. The last three values in the table do not make sense in the problem situation. Both the input and the output for this problem must be elements of the whole numbers.

g. Let $x = t$ and $y = 6t$. The graph appears below.

```
WINDOW FORMAT
Tmin=0
Tmax=700
Tstep=25
Xmin=-100
Xmax=700
Xscl=100
↓Ymin=-600
Ymax=4200
Yscl=600
```

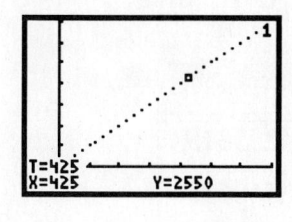

```
T=425
X=425    Y=2550
```

Even More Professors and Students

Professors	Students
425	2550
475	2050
525	3150
575	3450

h. The input values (x) represents the number of professors and the output values (y) represent the number of students.

7. a. $C(z) = 2z - 7$

z	$C(z)$
1	−5
2	−3
3	−1
4	1
5	3
6	5

b. and c.

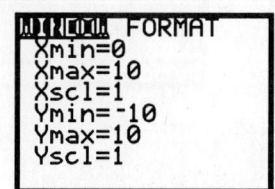

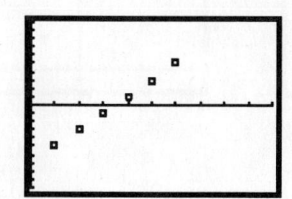

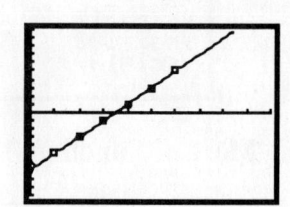

```
WINDOW FORMAT
Xmin=0
Xmax=10
Xscl=1
Ymin=-10
Ymax=10
Yscl=1
```

d. **More of Function C**

z	C
5.212766	3.4255319
5.4255319	3.8510638
2.3404255	−2.319149

e. $x = t$ and $y = 2t - 7$

f. Graph is in the same viewing window as b and c with Tstep equal to 0.5

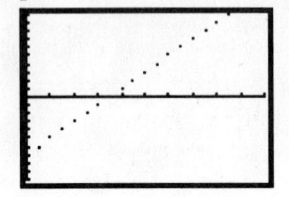

8. a. $A(b) = 3 + 2b^2$

b	$A(b)$
−5	53
−2	11
0	3
3	21
5	53
6	75

b. and c.

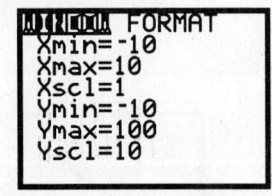

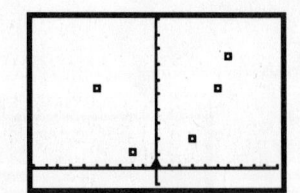

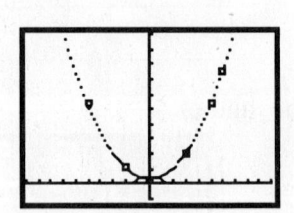

d. **More of Function A**

z	A
−3.829787	32.334541
−5.744681	69.002716
1.2765957	6.2593934

e. $x = t$ and $y = 3 + 2t^2$

f. Graph is in the same viewing window as b and c with Tstep equal to 0.5.

9. a. $L(r) = 3r^2 - 2r + 7$

r	$L(r)$
–4	63
–3	40
–1	12
0	7
2	15
4	47

b. and c.

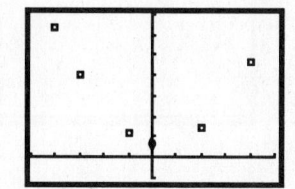

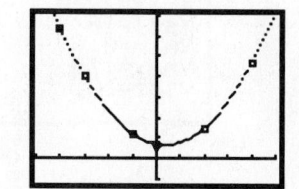

```
WINDOW FORMAT
Xmin=-5
Xmax=5
Xscl=1
Ymin=-10
Ymax=70
Yscl=10
```

d. **More of Function L**

z	L
–2.021277	23.29923
3.0851064	29.383431
0.31914894	6.6672703

e. $x = t$ and $y = 3t^2 - 2t + 7$

f. Graph is in the same viewing window as b and c with Tstep equal to 0.5.

10. From 0 to 10 seconds the elevator is waiting at Floor 1. From 10 seconds to 20 seconds, the elevator rises from Floor 1 to Floor 2. From 20 to 30 seconds, the elevator is waiting at Floor 2. From 30 to 40 seconds, the elevator has moved up from Floor 2 to just above Floor 3. From 40 to 50 seconds, the elevator has continued to rise to Floor 6. From 50 to 60 seconds and from 60 to 70 seconds, the elevator has remained at Floor 6.

11. a. The graph displays five points that have rapidly increasing y values for increasing x values.

   b.

   c. The output of the factorial function is undefined when the input is not a whole number. The domain of the factorial function is the set of whole numbers. These are two different ways of indicating what the possible inputs for the factorial function are.

12. a.

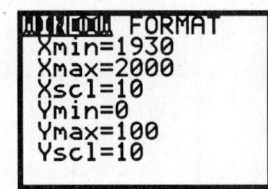

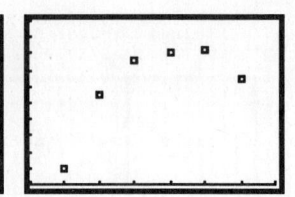

   b. The graph rises rapidly during the first twenty years indicating a rapid increase in the percentage of Americans with private health insurance. The graph levels off for the next twenty years indicating relatively little change in the percentage of Americans with private health insurance. Finally, the percentage covered drops in the last ten years.

   c. The first drop in percentage of Americans covered by private health insurance occurred during the decade from 1980 to 1990

13. a. **Width and Length of Rectangles**

Width (feet)	Length (feet)
0	10
2	8
4	6
6	4
8	2
10	0

b. $l = 10 - w$ where l represents the length and w represents the width.

c. The domain of the length function is the set of all real numbers which are greater than 0 and less than 10.

d. Any number between 0 and 10 would represent the width of one of the rectangles whose perimeter is 20 feet. So, any point on the part of the connected graph which is between the axes would be an ordered pair of values which would make sense in the problem situation.

14. The equation would be $2z - 7 = 35$. If we add 7 to both sides and simplify we get $2z = 42$. Finally, divide both sides by 2 to obtain the solution $z = 21$.

15. $3 + 2b^2 = 35 \qquad 2b^2 = 32 \qquad b^2 = 16 \qquad b = \pm 4$

16. a. $-(-2x + 7)$ \qquad b. $-(x^2 + 5x - 8)$

Section 4.6 Functions over the Integers

p. 334

2. a. **Table 4** $L(x)$, $OppL(x)$, and $AbsL(x)$

x	$L(x)$	$OppL(x)$	$AbsL(x)$
4	8	−8	8
−3.7	−7.4	7.4	7.4
0	0	0	0
−5	−10	10	10
15	30	−30	30

69

b. $OppL(x) = -2x$ and $AbsL(x) = |2x|$.

c. All graphs are in the standard viewing window.

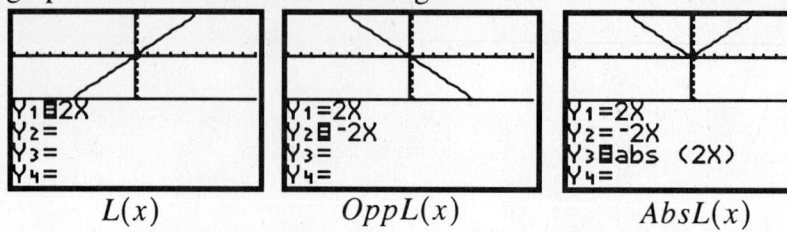

      $L(x)$ $OppL(x)$ $AbsL(x)$

d. The graph of $OppL(x)$ is obtained from the graph of $L(x)$ by reflection about the x–axis.

e. The graph of $AbsL(x)$ is obtained from the graph of $L(x)$ by reflecting all negative outputs about the x–axis. Non–negative outputs are not reflected.

3. Given $Q(x) = x^2 - 5x - 3$.

   a. **Table 5** $Q(x), OppQ(x),$ and $AbsQ(x)$

x	$Q(x)$	$OppQ(x)$	$AbsQ(x)$
–5	47	–47	47
–2	11	–11	11
0	–3	3	3
2	–9	9	9
4	–7	7	7

   b. $OppQ(x) = -(x^2 - 5x - 3) = -x^2 + 5x + 3$ and $AbsQ(x) = |x^2 - 5x - 3|$.

   c. All graphs are in the standard viewing window.

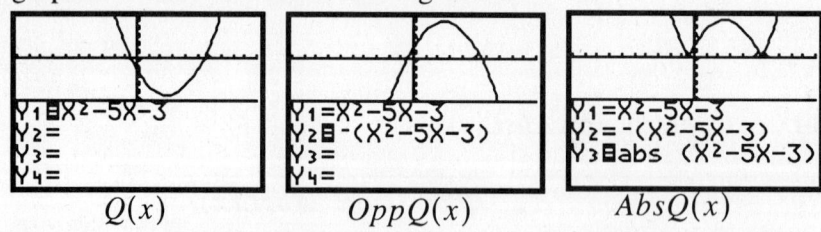

      $Q(x)$ $OppQ(x)$ $AbsQ(x)$

   d. The graph of $OppQ(x)$ is obtained from the graph of $Q(x)$ by reflection about the x–axis.

   e. The graph of $AbsQ(x)$ is obtained from the graph of $Q(x)$ by reflecting all negative outputs about the x–axis. Non–negative outputs are not reflected.

4. Let $G(x) = -(3x-5)$ and $H(x) = -3x+5$.

   a. **Table 6 $G(x)$ Versus $H(x)$**

x	$G(x)$	$H(x)$
−5	20	20
−2	11	11
0	5	5
2	−1	−1
4	−7	7

   b.

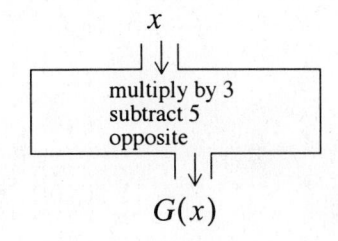

   x → [multiply by 3 / subtract 5 / opposite] → $G(x)$

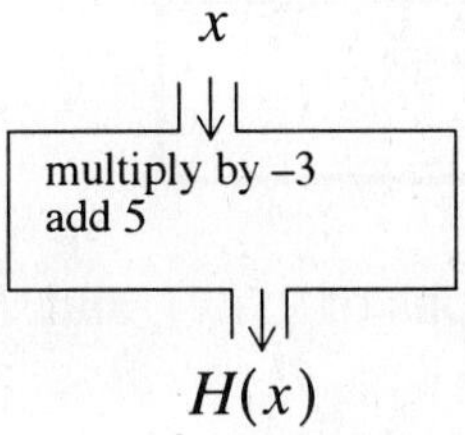

   x → [multiply by −3 / add 5] → $H(x)$

   c.

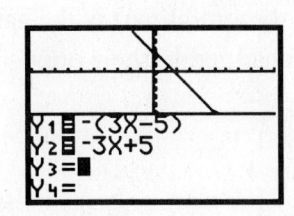

   ```
   Y1=-(3X-5)
   Y2=-3X+5
   Y3=
   Y4=
   ```

   d. The graphs of $G(x)$ and $H(x)$ are the same.

   e. The process represented by the algebraic representations are different. The function machine diagrams which appear in part b above show the two different processes. The two processes do have the same outputs for each input, however.

5. Let $G(x) = -(x^2 - 5x - 3)$ and $H(x) = -x^2 + 5x + 3$.

a. **Table 7 G(x) Versus H(x)**

x	G(x)	H(x)
-5	-47	-47
-2	-11	-11
0	3	3
2	9	9
4	7	7

b.
```
Y1=-(X²-5X-3)
Y2=-X²+5X+3
Y3=
Y4=
```

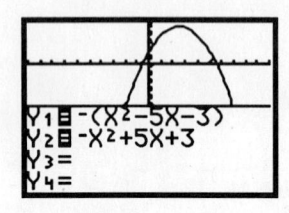

c. The graphs of $G(x)$ and $H(x)$ are the same.

d. In $G(x) = -(x^2 - 5x - 3)$ the polynomial part is evaluated first then the opposite of that number is determined. In $H(x) = -x^2 + 5x + 3$ the polynomial part is the same as the polynomial part in $G(x)$ except all the signs have been changed to their opposites.

6. Let $G(x) = -(3x^2 + 5x - 1)$ and $H(x) = -3x^2 - 5x + 1$.

a. **Table 8 G(x) Versus H(x)**

x	G(x)	H(x)
-5	-49	-49
-2	-1	-1
0	1	1
2	-21	-21
4	-67	-67

b.
```
Y1=-(3X²+5X-1)
Y2=-3X²-5X+1
Y3=
Y4=
```

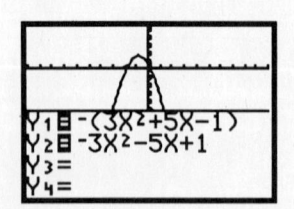

c. The graphs of $G(x)$ and $H(x)$ are the same.

d. The minus sign in front of the polynomial in $G(x) = -(3x^2 + 5x - 1)$ indicates taking the opposite of the polynomial last. In the expression which defines H, $H(x) = -3x^2 - 5x + 1$ each of the signs of the polynomial in $G(x)$ have been changed to their opposite.

7. a. The sign of x is negative.

    b. The sign of $-x$ is positive.

    c. The sign of $|x|$ is positive.

    d. Let $x = -43$ then $-x = -(-43) = 43$ and $|x| = -(-43) = 43$.

8. a. $f(x) = -(9 + 2x) = -9 - 2x$.

    b. $f(x) = -(3x^2 - 4x + 7) = -3x^2 + 4x - 7$

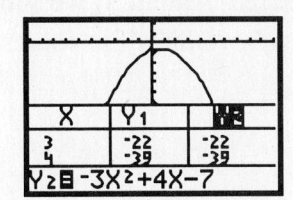

9. For $f(x) = -(9 + 2x)$, the output is negative for some values of x For example, if $x = -4$, then the output is -1.

   For $f(x) = -(3x^2 - 4x + 7)$, the output is always negative. This is proven by noting the graph is always in Quadrants III or IV. Either input in the given table produces a negative output.

10. a. $fun(x) = -(-13x + 4)$

    b. $joy(x) = -(x^2 + 2x - 3)$

73

11. A table is an excellent way to investigate this problem.

2–point shots	3–point shots
15	0
12	2
9	4
6	6
3	8
0	10

This problem could be investigated using an equation. If a represents the number of two–point shots made and b represents the number of three–point shots made, then the relationship between a and b is

$$2a + 3b = 30$$

where a and b are whole numbers.

12. Are the expressions $3 + 2m$ and $5m$ equivalent? They are not equivalent. If we substitute 0 for m then $3 + 2m = 3$ and $5m = 0$. Since the expressions have different values for a given value of the variable, they are not equivalent.

13. a. First look at a picture.

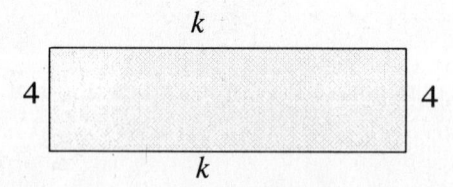

    The perimeter is the sum of the four sides which is $4 + 4 + k + k$ or $8 + 2k$.

    b. The area is the product of the length and the width. So an expression for the area is $4k$.

14. a. $x = -6$
    b. $x = -5$
    c. $x = \dfrac{4}{3}$
    d. $x = 23$ or $x = -5$
    e. No solutions.
    f. $x = 7$

Section 4.7 Review

p. 340

1. a. 8   b. a

 c. $7b$ d. $-3y + 2x$

 e. $-5x - 7$ f. $-x^2 + 3x - 11$

2. a. The first one is used to represent negative 2 the second one is subtraction with inputs –2 and 8.

 b. The second minus sign indicates the operation of taking the opposite of b. The first one represents subtraction with inputs a and $-b$.

 c. Both minus signs represent the operation of taking the opposite of the values of the variables.

 d. The first one is used to indicate the opposite of a. The second one means subtract with inputs $-a$ and b.

 e. The first and last minus signs are used to indicate taking the opposite the second one represents subtraction with inputs $-5a$ and $-2c$.

 f. The second minus sign means to take the opposite of a. The first one indicates subtraction with inputs 4 and $-a$.

 g. The second minus sign means to take the opposite of 4. The first one indicates subtraction with inputs a and -4.

3. b. For $a - (-b)$

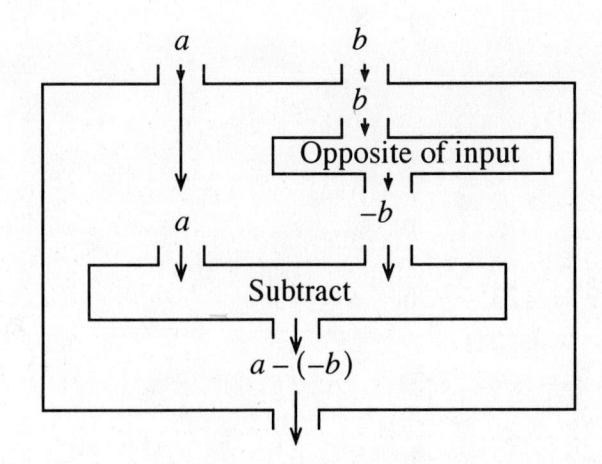

75

c. For $-a+(-b)$

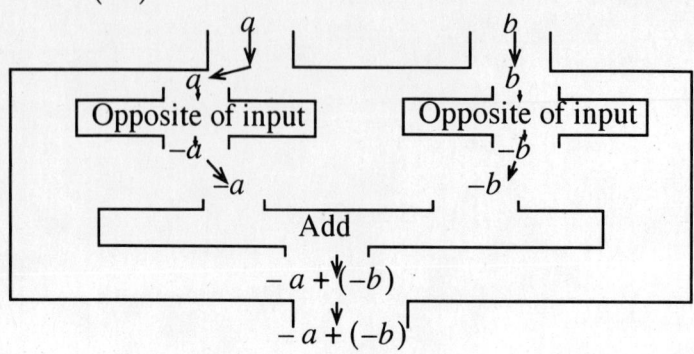

[diagram: input a → Opposite of input → $-a$; input b → Opposite of input → $-b$; Add → $-a+(-b)$]

d. For $-a-b$

4. a. $-3+(-7)$ b. $5+(-(8)) = 5+8$

[diagram: input a → Opposite of input → $-a$; input b; Subtract → $-a-b$]

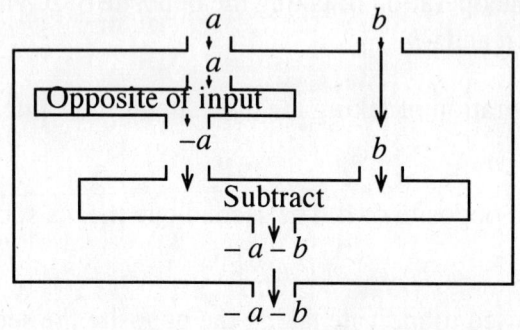

c. $-a+(-b)$ d. $-x+(-(y)) = -x+y$

5. a. -72 b. 1007
   c. 4 d. 4
   e. -19 f. -17
   g. -38 h. 20
   i. -66 j. 8
   k. 15 l. -15
   m. -11 n. -4
   o. -27 p. 6

6. a. $R(-2) = 7(-2)-9 = -23$ b. $x = -5$

7. $y(-5) = -5^2 - 4 \cdot 5 + 3 = -42$

8. a. $x = -7$ b. $b = 6$

76

c. $q = -26$ d. $x = 6$

e. $y = 0$ f. $x = 1$

g. $x = 5$ h. $x = 5$

i. $x = 9$ or $x = -5$ j. $x = 3$ or $x = -4$

k. $k = -2$

9. a. The temperature at 5:00 P.M. was 25^0 and the change in temperature since noon was $\Delta t = 20^0$

b. The temperature at 8:00 A.M. was -11^0 with a change of $\Delta t = 16^0$ from 8:00 A.M. until noon.

10. The change in miles using the mile markers was -88 miles. She had driven 88 miles.

11. a. $|a - 7|$ b. $|a - (-19)| = |a + 19|$

12. Xmin = -160, Xmax = 55, Xscl = 20, Ymin = -25, Ymax = 10, Yscl = 5.

13. Xmin = -12, Xmax = 15, Xscl = 3, Ymin = -150, Ymax = 105, Yscl = 15.

14. Xmin = -15, Xmax = 5, Xscl = 1, Ymin = -1097, Ymax = 343, Yscl = 100 would be one possible window. However, it would be a good practice to provide a larger viewing rectangle so the points at the ends of the graph are visible. The following choices would accomplish that: Xmin = -20, Xmax = 10, Xscl = 1, Ymin = -1300, Ymax = 350, Yscl = 100.

15. a. **Tutors and Students**

Tutors	Students
100	400
200	800
300	1200
400	1600
500	2000
600	2400

c.

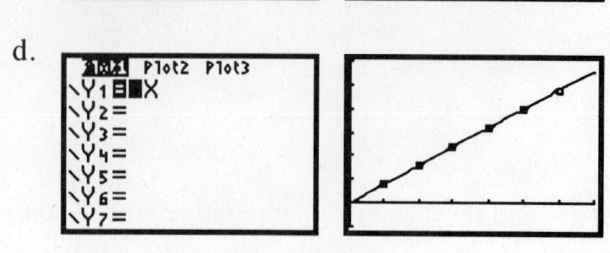

```
WINDOW
 Xmin=0
 Xmax=700
 Xscl=100
 Ymin=-500
 Ymax=3000
 Yscl=500
 Xres=1
```

1:L1,L2

X=600   Y=2400

d.
```
Plot1  Plot2  Plot3
\Y1=■X
\Y2=
\Y3=
\Y4=
\Y5=
\Y6=
\Y7=
```

e. **More Tutors and Students**

Tutors	Students
223.40426	893.61702
327.65957	1310.6383
424.46809	1697.8723
625.53191	2502.1277

f. The ordered pairs found by using the trace feature do not make sense in the problem situation since the numbers must be whole numbers.

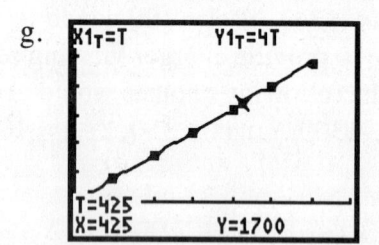

g.
```
X1T=T       Y1T=4T




T=425
X=425       Y=1700
```

Pairs That Make Sense

Tutors	Students
425	1700
450	1800
275	1100
150	600

h. The x-coordinates are the number of tutors and the y-coordinates are the corresponding number of students.

16. a. $-(4x-7) = -4x+7$ b. $-(2x^2+5x-11) = -2x^2-5x+11$

78

These equations are true because of the general principle that the operation of taking opposites distributes over addition and subtraction.

To check our answers numerically we enter the left side of one of the equations as Y1 and the right side of the equation as Y2 in our Y= menu. Then produce a table where the calculated values are produced from both expressions. The outputs should be the same for each input.

17. a. $5x + 9 = -(-5x - 9)$ b. $3t^2 - 5t + 7 = -(-3t^2 + 5t - 7)$

18. a. positive   b. negative

     c. positive   d. If $p = 3$ then $-p = -3$ and $|p| = 3$

19. a. negative   b. positive

     c. positive   d. If $n = -5$ then $-n = 5$ and $|n| = 5$.

20. a. x^{12} b. x^{35}

     c. $6x^{13}y^7$ d. -1

     e. This is simplified.   f. 0

     g. 1   h. c^{x+y}

     i. c^{xy} j. There is no property that can be used to simplify this expression.

21. a. $(5x^3 + 7x^2 - 8x + 2) - (5x - 4x^3 + 7 + x^2) =$

     $5x^3 + 7x^2 - 8x + 2 - 5x + 4x^3 - 7 - x^2 = 9x^3 + 6x^2 - 13x - 5$

     b. $35x^4 - 26x^3 - 42x^2 + 37x - 6$

     c. $16x^2 - 9y^2$ d. $16x^2 - 24xy + 9y^2$

     e. $x^3 - 12x^2 + 48x - 64$

CHAPTER 5

Rational Numbers: Further Expansion of a Mathematical System

Section 5.1 Rates of Change

p. 358

2. a. **Table 4 Car Rental Data**

Finite Difference in Input Δm	Miles Driven m	Total Cost C ($)	Finite Difference in Output ΔC	$\dfrac{\Delta C}{\Delta m}$
96	15	37.55	16.32	.17
−19	111	53.87	−3.23	.17
200	92	50.64	34	.17
−142	292	84.64	−24.14	.17
57	150	60.5	9.69	.17
−164	207	70.19	−27.88	.17
	43	42.31		

b. They are the same.

c. The ratio of the change in the outputs to the change in the inputs is the same even if the inputs do not change by a constant amount. This is because in our function for determining the cost when the number of miles is changed the cost is increased by 0.17 times that change or the cost per mile.

3. a. **Concert Receipts**

Number of tickets sold (n)	Total receipts in dollars (R)
200	3800
400	7600
600	11400
800	15200
1000	19000

80

b. $R(n) = 19n$

c. $slope = \dfrac{\Delta R}{\Delta n} = 19$.

d. The slope represents the rate of change of the receipts per each ticket sold. In this case, for each ticket sold, the total receipts increase by $19.

e. The finite differences of the outputs will not be constant but the ratio of the change in output to the change in input between any two input-output pairs will be 19.

4. All would be labeled significant barriers except DE, which is fully accessible.

5. a. Graph (a) Shows their distance they traveled at each time during the first trip. Graph (b) shows the speed they were traveling at each time during the second trip.

b. In graph (a) point C shows that they were at the same distance from the starting point at the end of the trip. In graph (b) point C shows that at the time which corresponds to point C they were traveling at the same speed.

c. The total distance each traveled is the second coordinate of the point C. They both traveled the same distance on this trip. Bev covered more of the total distance in the beginning of the trip than Al did.

d. Bev was going faster than Al for the entire trip. So, Al's total distance would be less than Bev's.

e. Bev continued to increase her speed during the entire trip. This is because the graph continues to increase as a point on the graph moves from left to right.

6. a. The second lot is longer. One uses an additive comparison between the lenghts to answer this question.

b. The second lot has more area. After multiplying the length and the width of each lot to get the areas we use and additive comparison to tell which is bigger.

c. The 80 feet by 55 feet lot would be closer to a square because $\dfrac{55}{80}$ is slightly closer to one than $\dfrac{75}{110}$. The closer the rectangle is to a square the closer the ratio of the length to the width would be to 1. Here we use a multiplicative comparison.

7. a. $\dfrac{\Delta(mileage)}{\Delta(speed)} = \dfrac{14-17}{60-45} = -\dfrac{1}{5}$. The average rate of change of gas mileage with respect to speed is $-\dfrac{1}{5}$. This means that for every five m.p.h. increase in speed, the gas mileage decreases by one mile per gallon.

b. At 50 m.p.h., the gas mileage is 16 miles per gallon. At 45 m.p.h., the mileage was 17 miles per gallon. By increasing the speed by five m.p.h., we must decrease the mileage by 1 mile per gallon.

At 70 m.p.h., the mileage is 12 miles per gallon. At 60 m.p.h., the mileage was 14 miles per gallon. By increasing the speed by 10 m.p.h., we must decrease the mileage by 2 miles per gallon.

c. A negative rate of change means that as the input increases the output decreases. In this problem, as the speed (input) increases, the gas mileage (output) decreases.

8. a. **Table 6 Change in Distance Fallen of the Stale Corned Beef Sandwich**

Change in time (seconds)	Change in distance fallen (feet)
1 to 2	48
2 to 3	80
3 to 4	112
4 to 5	144
5 to 6	176
6 to 7	208
7 to 8	240
8 to 9	272
9 to 10	304

b. The rate of change in distance fallen (height) with respect to time is varying. As time changes by one second, the distance fallen each second increases by 32 feet.

c. The average rate of change in distance fallen with respect to time represents the ratio of the total distance fallen to the total elapsed time.

d. $\dfrac{\Delta d}{\Delta t} = \dfrac{1600 - 16}{10 - 1} = \dfrac{1584}{8} = 176$ feet per second. The poor sandwich has fallen at an average rate of 176 feet per second. Here we actually used Table 5 instead of Table 6.

e. The sandwich moves the fastest over the time interval 9–10. In this interval, it has fallen 304 feet. The sandwich moves the slowest over the time interval 0–1. In this interval, it has fallen 16 feet.

9.   Let d represent distance in miles and t represent time in hours.

   a. 10–12: $\dfrac{\Delta d}{\Delta t} = \dfrac{78334 - 78239}{12 - 10} = \dfrac{95}{2} = 47.5$ m.p.h.

   12–1:15: $\dfrac{\Delta d}{\Delta t} = \dfrac{78334 - 78334}{1\frac{1}{4}} = 0$ m.p.h.

   1:15–2:30: $\dfrac{\Delta d}{\Delta t} = \dfrac{78383 - 78334}{1\frac{1}{4}} = \dfrac{49}{\frac{5}{4}} = 49\left(\dfrac{4}{5}\right) = 39.2$ m.p.h.

   2:35–5:10: $\dfrac{\Delta d}{\Delta t} = \dfrac{78489 - 78383}{2\frac{7}{12}} = \dfrac{106}{\frac{29}{12}} = 106\left(\dfrac{12}{29}\right) \approx 44$ m.p.h.

   b. $\dfrac{\Delta d}{\Delta t} = \dfrac{78489 - 78239}{7\frac{1}{6}} = \dfrac{250}{\frac{43}{6}} = 50\left(\dfrac{6}{43}\right) \approx 34.9$ m.p.h. The average speed over the entire trip was about 35 m.p.h.

10. A is used as a label while a is used as a variable representing altitude.

11.   a. $slope = \dfrac{4 - 11}{6 - 3} = -\dfrac{7}{3}$.

   b. $slope = \dfrac{-6 - 3}{4 - (-5)} = \dfrac{-9}{9} = -1$.

   c. $slope = \dfrac{-4 - (-8)}{-7 - 0} = -\dfrac{4}{7}$.

   d. $slope = \dfrac{-3 - (-3)}{15 - 2} = 0$.

   e. $slope = \dfrac{-3 - 2}{4 - 4} =$ undefined.

12.   a. 3   b. –2

   c. 5   d. –4

13. a is the rate of change of the outputs to inputs.

14. a. $\dfrac{y(4)-y(1)}{4-1} = \dfrac{(2\cdot 4+3)-(2\cdot 1+3)}{3} = \dfrac{11-5}{3} = \dfrac{6}{3} = 2$ for $y(x) = 2x+3$.

$\dfrac{y(4)-y(1)}{4-1} = \dfrac{(4^2+5)-(1^2+5)}{3} = \dfrac{21-6}{3} = \dfrac{15}{3} = 5$ for $y(x) = x^2+5$.

b. $\dfrac{y(5)-y(3)}{5-3} = \dfrac{(2\cdot 5+3)-(2\cdot 3+3)}{2} = \dfrac{13-9}{2} = \dfrac{4}{2} = 2$ for $y(x) = 2x+3$.

$\dfrac{y(5)-y(3)}{5-3} = \dfrac{(5^2+5)-(3^2+5)}{2} = \dfrac{30-14}{2} = \dfrac{16}{2} = 8$ for $y(x) = x^2+5$.

c. $\dfrac{y(9)-y(4)}{9-4} = \dfrac{(2\cdot 9+3)-(2\cdot 4+3)}{5} = \dfrac{21-11}{5} = \dfrac{10}{5} = 2$ for $y(x) = 2x+3$.

$\dfrac{y(9)-y(4)}{9-4} = \dfrac{(9^2+5)-(4^2+5)}{5} = \dfrac{86-21}{5} = \dfrac{65}{5} = 13$ for $y(x) = x^2+5$.

15. The linear function $y(x) = 2x+3$ has a constant rate of change of outputs to inputs. In the linear function the variable is raised to the first power only. In the quadratic function, $y(x) = x^2+5$, the variable is squared.

16. a.

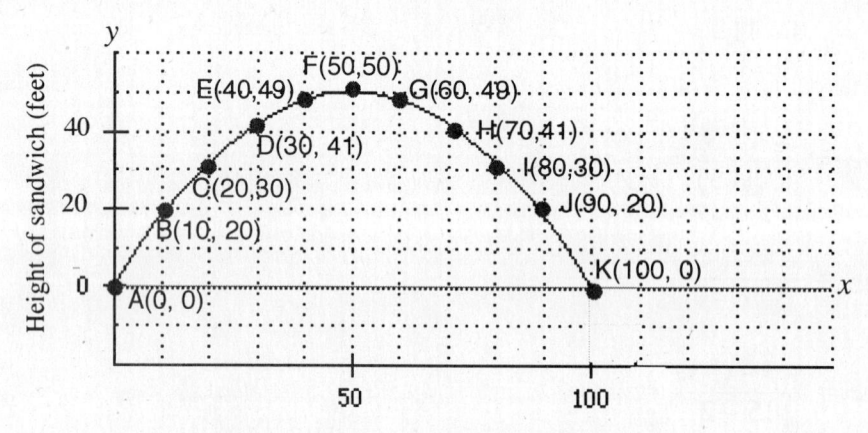

b. Let a represent altitude in feet and d represent horizontal distance traveled in feet.

AB: $\dfrac{\Delta a}{\Delta d} = \dfrac{20-0}{10-0} = \dfrac{2}{1} = 2$. For every 1 foot change horizontally, the sandwich's height has increased by 2 feet.

BC: $\dfrac{\Delta a}{\Delta d} = \dfrac{30-20}{20-10} = 1$. For every 1 foot change horizontally, the sandwich's height has increased by 1 foot.

CD: $\dfrac{\Delta a}{\Delta d} = \dfrac{41-30}{30-20} = \dfrac{9}{10}$. For every 1 foot change horizontally, the sandwich's height has increased by $\dfrac{9}{10}$ of a foot.

DE: $\dfrac{\Delta a}{\Delta d} = \dfrac{49-41}{40-30} = \dfrac{8}{10} = \dfrac{4}{5}$. For every 1 foot change horizontally, the sandwich's height has increased by $\dfrac{4}{5}$ of a foot.

EF: $\dfrac{\Delta a}{\Delta d} = \dfrac{50-49}{50-40} = \dfrac{1}{10}$. For every 1 foot change horizontally, the sandwich's height has increased by $\dfrac{1}{10}$ of a foot.

FG: $\dfrac{\Delta a}{\Delta d} = \dfrac{49-50}{60-50} = -\dfrac{1}{10}$. For every 1 foot change horizontally, the sandwich's height has decreased by $\dfrac{1}{10}$ of a foot.

GH: $\dfrac{\Delta a}{\Delta d} = \dfrac{41-49}{70-60} = -\dfrac{8}{10} = -\dfrac{4}{5}$. For every 1 foot change horizontally, the sandwich's height has decreased by $\dfrac{4}{5}$ of a foot.

HI: $\dfrac{\Delta a}{\Delta d} = \dfrac{30-41}{80-70} = -\dfrac{9}{10}$. For every 1 foot change horizontally, the sandwich's height has decreased by $\dfrac{9}{10}$ of a foot.

IJ: $\dfrac{\Delta a}{\Delta d} = \dfrac{20-30}{90-80} = -1$. For every 1 foot change horizontally, the sandwich's height has decreased by 1 foot.

JK: $\dfrac{\Delta a}{\Delta d} = \dfrac{0-20}{100-90} = -\dfrac{20}{10} = -2$. For every 1 foot change horizontally, the sandwich's height has decreased by 2 feet.

c. The sandwich is rising when the rate of change is positive and the sandwich is falling when the rate of change is negative.

d. The sandwich is at its highest point at about (50, 51).

e. The rate of change of altitude with respect to the horizontal distance traveled is about 0 near the highest point.

f. Before the sandwich reaches its highest point, the rates of change are positive. After the sandwich reaches its highest point, the rate of change are negative.

g. The sandwich is at its highest point and is, for an instant, neither gaining nor losing altitude.

17. The traveler whose motion is graphed started 1000 feet from some fixed reference point. In the first 3 hours he traveled another 1000 feet. From hour 3 to hour 4 he traveled 2000 feet and is 4000 feet from the fixed reference point. From point E to point F the traveler moved closer to the fixed reference point by 1000 feet in two hours. The motions between the other points are similar.

It is legitimate to use horizontal distance traveled as the label on the vertical axis.

18. a. After five years the profits for CDs-R-Us is at $60,000 while the profits for Super Dog Burger Bits is at $300,000.

b. The rate of change of profits per year for CDS-R-Us is $20,000 per year computed using the following ratio: $\frac{60-20}{5} = \frac{40}{5} = 20$. The rate of change in profits per year for Super Dog Burger Bits is $40,000 per year computed using the ratio $\frac{300-100}{5} = \frac{200}{5} = 40$.

Section 5.2 Rational Numbers and Proportional Reasoning

p.380

2. a. $\frac{32}{20} = \frac{8}{5}$ apples per oranges b. $\frac{33}{51} = \frac{11}{17}$ men per women

   c. $\frac{12}{8} = \frac{3}{2}$ dogs per cats d. $\frac{190}{4} = \frac{95}{2}$ miles per hour

   e. $\frac{36}{8} = \frac{9}{2}$ pounds per square inch f. $\frac{16}{9}$ soda per ice tea; is reduced.

3. a. $\frac{1}{6}$ b. $-\frac{1}{5}$

4. No since there are an infinite number of rational numbers between every pair of rational numbers.

5. $\frac{a}{b}$ can be interpreted as a rational number, as division of a by b., as the answer to the division problem a divided by b, and as the ratio of a to b.

    The reason that b cannot be zero in $\frac{a}{b}$ is that the product $b\left(\frac{a}{b}\right)$ must be a. But if b was zero this product would always be zero, since multiplying by zero yields zero.

6. a. **Table 5: More Reciprocal Exploration**

Number	Reciprocal
8	.125
7	.142857...
6	.1666666...
5	.2
4	.25
3	.333333...
2	.5

    b. It gets closer to the reciprocal of 1 which is 1.

7. a. Cannot be simplified since the numerator and denominator have no common factor other than 1.

    b. $\frac{3x}{2x} = \frac{3}{2}$ c. $\frac{(x+2)(x-3)}{(x-3)(x+4)} = \frac{x+2}{x+4}$

    d. $\frac{2x(x+3)}{(x-3)4x} = \frac{2(x+3)}{4(x-3)} = \frac{x+3}{2(x-3)}$

8. a. All numbers except -2.   b. All numbers except 0.

    c. All numbers except 3 and -4.   d. All numbers except 3 and -4.

    e. All numbers except 3 and 0.

9. a. $\frac{F}{P} = \frac{2}{7}$ b. $\frac{2}{14}, \frac{6}{21}, \frac{8}{28}, \frac{10}{35}, \frac{12}{42}$

c. **Table 6: Full-time versus part-time students**

P	F
7	2
14	4
21	6
28	8
35	10

d. $F(P) = \frac{2}{7}P$

e.

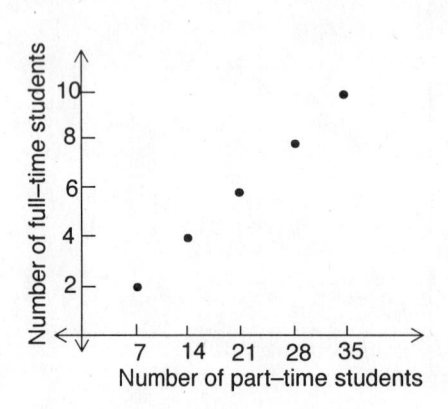

10. Tom is getting the flour at $0.75 per pound, found by changing $\frac{3}{4}$ to a decimal, while Sue is able to buy the flour at little more than $0.71 per pound, found by changing $\frac{5}{7}$ to a decimal. Sue would be getting a better deal.

11. a. If we let w stand for the number of women in the town and m stand for the number of men in the town then since two-thirds of the women are married to three-fifths of the men the equation $\frac{2}{3}w = \frac{3}{5}m$ must be satisfied. If we multiply both sides of this equation by $\frac{5}{3}$ we convert the equation into the form $m = \frac{5}{3} \cdot \frac{2}{3}w = \frac{10}{9}w$. The least number of people living in the town would be 9 women and 10 men because if we multiply $\frac{10}{9}$ by anything smaller than 9 we won't get a whole number.

b. The values of *w* that would produce whole numbers would be counting number multiples of 9. Some of the possible numbers of people living in the town are shown in the following table.

Possible Numbers of People in the Town

Women	Men
18	20
45	50
108	120

12. a. **Original price versus sale price**

Original price P	Sale Price S
10	7.30
20	14.60
30	21.90
40	29.20
50	36.50

b. $S(P) = P - 0.27P$ or $S(P) = 0.73P$

c.

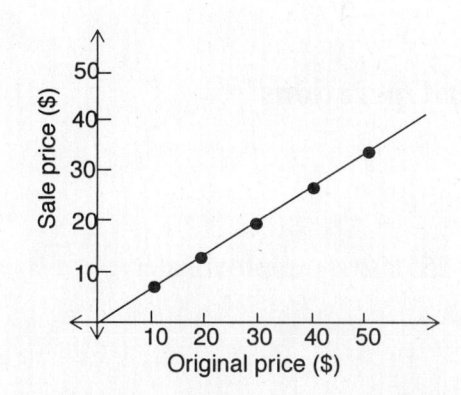

13. a. The Tasty & Juicy has 0.3125 ounces of concentrate per ounce of water while the Delicious Fruit Juice has 0.3 ounces of concentrate per ounce of water. The Tasty & Juicy is stronger.

b. It would take $11\frac{7}{13}$ ounces. Solve this algebraically by letting *c* represent the required amount of concentrate and *w* the necessary amount of water. We know that $\frac{c}{w} = \frac{3}{10}$

which can be rewritten to see that $w = \frac{10}{3}c$. And since the water and the concentrate together must add to 50 ounces w and c must satisfy $w + c = 50$. Replacing w with $\frac{10}{3}c$ gives us the equation $\frac{10}{3}c + c = 50$ to solve. This equation is equivalent to $\frac{13}{3}c = 50$ or $c = \frac{3}{13} \cdot 50 = \frac{150}{13} = 11\frac{7}{13}$

    c. $11\frac{19}{21}$ ounces would be required.

14. a. **Price paid by store versus marked price**

Change in Price Paid by Store	Price Paid by Store ($)	Marked Price ($)	Change in Marked Price
5	10	14.5	7.25
5	15	21.75	7.25
	20	29	

    b. The rate of change of the marked price to the price the price the store paid is $\frac{7.25}{5} = 1.45$.

    c. If $M(p)$ is the marked price and p is the price the store paid then $M(p) = 1.45p$.

Section 5.3 Investigating Rational Number Operations

p. 400

2. a. $-\frac{3}{2}$ b. 0.25510204 approximately

    c. $\frac{11}{86}$ d. $\frac{1}{5x}$

    e. $-\frac{1}{3y}$ f. $\frac{1}{x+y}$

3. a. $5\left(\frac{1}{17}\right)$ b. $-3(11)$

c. $-2.3\left(\dfrac{6}{5}\right)$ d. $\dfrac{4}{7}\left(-\dfrac{1}{2}\right)$

e. $-9.2\left(-\dfrac{5}{11}\right)$ f. $x\left(\dfrac{1}{3y}\right)$

g. $5t(q)$ h. $\dfrac{3}{k}\left(\dfrac{c}{d}\right)$

i. $\dfrac{3x+4}{x-5} \cdot \dfrac{x+5}{9x}$ j. $\dfrac{y-2}{2x-6} \cdot \dfrac{4x+8}{2y+1}$

4. a. To add two rational numbers in fractional form, determine the common denominator and expand each fraction so all fractions have the same denominator. Then add the numerators and keep the denominator. Reduce if possible.

   b. To add two rational numbers in decimal form, align the decimal points and add digits with like place value.

5. a. The procedure is identical to that stated in the answer to 4a except that we find the difference, rather than the sum, of the numerators.

   b. The procedure is identical to that stated in the answer to 4b except that we subtract digits, rather than add, with like place value.

6. a. To multiply two rational numbers in fractional form, multiply the numerators to find the numerator of the product and multiply the denominators to find the denominator of the product. Reduce, if possible.

   b. To multiply two rational numbers in decimal form, multiply the numbers without the decimal points. The number of decimal places in the product equals the sum of the number of decimal places in the two factors.

7. a. $\left(\dfrac{7}{8}\right)\left(\dfrac{2}{3}\right) = \left(\dfrac{7}{4}\right)\left(\dfrac{1}{3}\right) = \dfrac{7}{12}$

   b. $\dfrac{7}{5} \div \left(-\dfrac{42}{17}\right) = \dfrac{7}{5}\left(-\dfrac{17}{42}\right) = \dfrac{1}{5}\left(-\dfrac{17}{6}\right) = -\dfrac{17}{30}$

   c. $-\dfrac{5}{8}+\dfrac{1}{12} = -\dfrac{15}{24}+\dfrac{2}{24} = -\dfrac{13}{24}$

   d. $\dfrac{9}{10}-\left(-\dfrac{3}{4}\right) = \dfrac{9}{10}+\dfrac{3}{4} = \dfrac{18}{20}+\dfrac{15}{20} = \dfrac{33}{20}$

   e. $\left(-\dfrac{2}{9}\right)\left(\dfrac{11}{5}\right) = -\dfrac{22}{45}$

f. $-7\frac{3}{4} - 4\frac{2}{3} = -\frac{31}{4} + \left(-\frac{14}{3}\right) = -\frac{93}{12} + \left(-\frac{56}{12}\right) = -\frac{149}{12} = -12\frac{5}{12}$

g. $\frac{5}{8} + \left(-\frac{7}{10}\right) = \frac{25}{40} + \left(-\frac{28}{40}\right) = -\frac{3}{40}$

h. $\left(-\frac{13}{12}\right) \div \left(-\frac{5}{9}\right) = -\frac{13}{12}\left(-\frac{9}{5}\right) = -\frac{13}{4}\left(-\frac{3}{5}\right) = \frac{39}{20} = 1\frac{19}{20}$

i. $\frac{2}{9} + \left(-\frac{7}{12}\right) = \frac{8}{36} + \left(-\frac{21}{36}\right) = -\frac{13}{36}$

j. $\frac{7}{9} - \left(\frac{5}{3}\right)\left(\frac{3}{7}\right) = \frac{7}{9} - \left(\frac{5}{1}\right)\left(\frac{1}{7}\right) = \frac{7}{9} - \left(\frac{5}{7}\right) = \frac{7}{9} + \frac{5}{7} = \frac{49}{63} + \frac{45}{63} = \frac{94}{63} = 1\frac{31}{63}$

8. a. $\dfrac{2x+7}{x+5}$ b. $\dfrac{3y-11}{2y-1}$

c. $\dfrac{3x-2}{4x+7} + \dfrac{x+1}{4x+7} = \dfrac{4x-1}{4x+7}$ d. $\dfrac{2t+9}{3t-2}$

e. $\dfrac{8}{10x} + \dfrac{x^2}{10x} = \dfrac{8+x^2}{10x}$ f. $\dfrac{9x^2}{24x} - \dfrac{28}{24x} = \dfrac{9x^2 - 28}{24x}$

g. $\dfrac{3(x+3)}{(x+2)(x+3)} + \dfrac{4(x+2)}{(x+2)(x+3)} = \dfrac{3x+9+4x+8}{(x+2)(x+3)} = \dfrac{7x+17}{(x+2)(x+3)}$

h. $\dfrac{(3x-1)(x-4)}{(2x+5)(x-4)} - \dfrac{(2x+3)(2x+5)}{(2x+5)(x-4)} = \dfrac{(3x-13x+4) - (4x^2+16x+15)}{(2x+5)(x-4)} =$

$\dfrac{3x^2 - 13x + 4 - 4x^2 - 16x - 15}{(2x+5)(x-4)} = \dfrac{-x^2 - 29x - 11}{(2x+5)(x-4)}$

9. a. $x = \dfrac{22}{4} = \dfrac{11}{2}$ b. $y = -\dfrac{13}{3}$

c. $x = -\dfrac{22}{6} = -\dfrac{11}{3}$ d. $y = \dfrac{4.8 - 7.63}{3.41} \approx -0.8299$

e. $5x = 3 - \dfrac{2}{7} = \dfrac{19}{7}$
$x = \dfrac{19}{7} \div 5 = \dfrac{19}{7} \cdot \dfrac{1}{5} = \dfrac{19}{35}$

f. $\dfrac{3}{2}x = 4 + \dfrac{2}{5} = \dfrac{22}{5}$
$x = \dfrac{2}{3} \cdot \dfrac{22}{5} = \dfrac{44}{15}$

g. $\frac{5}{6}x = \frac{2}{3} + \frac{7}{9} = \frac{13}{9}$

$x = \frac{6}{5} \cdot \frac{13}{9} = \frac{2}{5} \cdot \frac{13}{3} = \frac{26}{15}$

h. $x = -\frac{5}{6} - \frac{7}{4} = -\frac{31}{12}$

10. a. $\left(\frac{1}{4} + \frac{1}{3}\right) \div 2 = \left(\frac{3}{12} + \frac{4}{12}\right) \div 2 = \frac{7}{12} \div 2 = \frac{7}{12}\left(\frac{1}{2}\right) = \frac{7}{24}$

b. $\left(\frac{1}{8} + \frac{1}{7}\right) \div 2 = \left(\frac{7}{56} + \frac{8}{56}\right) \div 2 = \frac{15}{56} \div 2 = \frac{15}{56}\left(\frac{1}{2}\right) = \frac{15}{112}$

c. There is a rational number between any two rational numbers because the average of the two numbers is a rational number between the two given rational numbers.

d. There is no integer between two consecutive integers.

11. Let $T(r) = 5r - 2$.

a. $T\left(\frac{3}{5}\right) = 5\left(\frac{3}{5}\right) - 2 = 3 - 2 = 1$. Here we are evaluating the function.

b. We are solving the equation $5r - 2 = 16$ to get $r = \frac{18}{5}$.

12. If $P(x) = 3x^2 - x - 4$ then

$P\left(-\frac{2}{7}\right) = 3\left(-\frac{2}{7}\right)^2 - \left(-\frac{2}{7}\right) - 4 = 3\left(\frac{4}{49}\right) + \frac{2}{7} - 4 = \frac{12}{49} + \frac{2}{7} - 4 =$

$\frac{12}{49} + \frac{14}{49} - \frac{196}{49} = \frac{26}{49} - \frac{196}{49} = -\frac{170}{49}$. We are evaluating the function.

13. a. **Table 4: More Reciprocal Exploration**

Number	Reciprocal
8	$\frac{1}{8} = 0.125$
7	$\frac{1}{7} \approx 0.143$

Number	Reciprocal
6	$\frac{1}{6} \approx 0.167$
5	$\frac{1}{5} = 0.2$
4	$\frac{1}{4} = 0.25$
3	$\frac{1}{3} \approx 0.333$
2	$\frac{1}{2} = 0.5$

    b. The value of the reciprocal is increasing toward 1 as the number decreases toward 1.

14. The width must be $\frac{2}{3}$ of the length so if w stands for the width and l stands for the length the equation $w = \frac{2}{3}l$ must be satisfied. For our width of $23\frac{7}{8}$ inches we must solve the equation $23\frac{7}{8} = \frac{2}{3}l$. Multiply both sides by $\frac{3}{2}$ to get $l = \frac{3}{2} \cdot 23\frac{7}{8} = \frac{3}{2} \cdot \frac{191}{8} = \frac{573}{16} = 35\frac{13}{16}$ inches.

15. $7\frac{3}{4}$ miles is $3\frac{1}{2}$ times as far as Bill can walk in the same amount of time. To find how far Bill walked we divide $7\frac{3}{4}$ miles by $3\frac{1}{2}$. $\left(7\frac{3}{4}\right) \div \left(3\frac{1}{2}\right) = \frac{31}{4} \div \frac{7}{2} = \frac{31}{4} \cdot \frac{2}{7} = \frac{31}{2} \cdot \frac{1}{7} = \frac{31}{14}$

So, Bill walked $2\frac{3}{14}$ miles.

16. a. The integers are closed under the operations of addition, subtraction, and multiplication.

    b. The rational numbers are closed under the operations of addition, subtraction, multiplication, and division.

Section 5.4 Reciprocal and Power Functions

p. 416

2. a. For example, the problems that dealt with the triangular numbers, car rentals, students and professors, concert tickets, and the Bulls T–shirts all contained functions which were discussed.

b. See previous sections in text dependent on the functions you chose.

3. $Power(x) = x^n$ where n is odd.

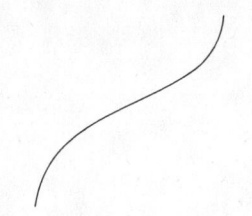

4. $Power(x) = x^n$ where n is even.

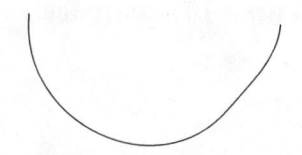

5. a. The amount you must study as a function of the number of credit hours you take. The input is the number of credit hours and the output is the amount of study time.

   b. The amount of free time that you have as a function of the number of credit hours you take. The input is the number of credit hours and the output is the amount of free time.

   c. The area of a cross section of a tree trunk is a function of the diameter of the tree trunk. The input is the diameter and the output is the area of the cross section.

   d. The average daily temperature in Chicago as a function of the date. The input is the date and the output is the average daily temperature.

   e. The cost of mailing a letter as a function of the letter's weight. The input is the weight of the letter and the output is the postal charge.

6. This problem is not possible since it will take 4 hours to go 120 miles if the average speed is 30 m.p.h. The total trip cannot then be made in 4 hours.

7. a. x^3 b. x^4
   c. x^5 d. x^5

8. x^{m+n}.

9. a. x^3 b. x^6
   c. x^8 d. x^6

10. x^{mn}

11. a. x b. x^2

　　c. x^3 d. x

　　e. x^2 f. x^3

12. x^{m-n}

　　Property: If $m > n$ then $\dfrac{x^m}{x^n} = x^{m-n}$.

13. a. **Table 3: Width versus length of a rectangle with area 10 square feet.**

Width	Length	Area
1	10	10
2	5	10
4	2.5	10
5	2	10
10	1	10
w	$\dfrac{10}{w}$	$w \cdot \dfrac{10}{w} = 10$

　　b. Find the length by dividing the area, 10, by the width.

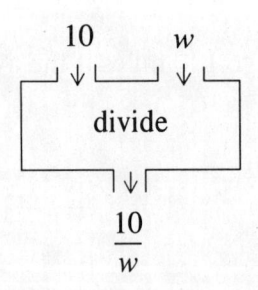

　　c.

　　　　　　10　　　w
　　　　　　↓　　　↓
　　　　　[divide]
　　　　　　　↓
　　　　　　$\dfrac{10}{w}$

　　d. $l(w) = \dfrac{10}{w}$

96

e.

```
WINDOW FORMAT
Xmin=0
Xmax=10
Xscl=1
Ymin=0
Ymax=10
Yscl=1
```

f. Some examples appear in the table below.

Width	Length
5.106383	1.958333
1.3829787	7.2307692
7.3404255	1.3623188

g. If the length is 3.8 feet then the width is about 2.63 feet.

14. a. $x = \dfrac{18}{5}$ b. $y = \dfrac{28}{3}$

c. $\dfrac{4}{5}t = \dfrac{2}{3} + 3 = \dfrac{11}{3}$
   $t = \dfrac{5}{4} \cdot \dfrac{11}{3} = \dfrac{55}{12}$

d. $y = -\dfrac{1}{9}$

15. a.

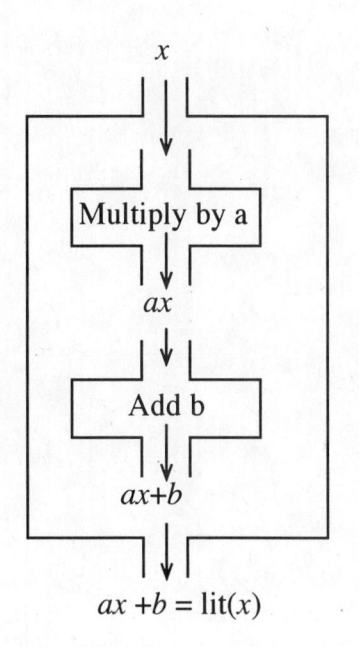

```
        x
        ↓
   ┌─────────────┐
   │ Multiply by a│
   └─────────────┘
        ↓
        ax
        ↓
   ┌─────────────┐
   │    Add b    │
   └─────────────┘
        ↓
       ax+b
        ↓
   ax + b = lit(x)
```
lit

b.
```
        x
        ↑
        x
        ↑
   ┌─────────────┐
   │ Divide by a │
   └─────────────┘
        ↑
        ax
        ↑
   ┌─────────────┐
   │  Subtract b │
   └─────────────┘
        ↑
       ax+b
        ↑
   ax + b = lit(x)
```
reverselit

16. a. $w = \dfrac{A}{l}$ b. $2l = P - 2w$
 $l = \dfrac{P - 2w}{2}$

c. $\dfrac{9}{5}C = F - 32$
 $C = \dfrac{5}{9}(F - 32)$

Section 5.5 Integer Exponents

p. 427

2. a. $8^{-2} = \dfrac{1}{8^2} = \dfrac{1}{64}$ b. $7^{-1} = \dfrac{1}{7^1} = \dfrac{1}{7}$

   c. $k^{-5} = \dfrac{1}{k^5}$

3. a. $\dfrac{k^7}{k^2} = k^{7-2} = k^5$ b. $\dfrac{t^4}{t^{11}} = t^{4-11} = t^{-7} = \dfrac{1}{t^7}$

   c. $\dfrac{z^5}{z} = z^{5-1} = z^4$ d. $\dfrac{x}{x^3} = x^{1-3} = x^{-2} = \dfrac{1}{x^2}$

   e. $\dfrac{x^4}{y^2}$ has different bases f. $\dfrac{y^{17}}{y^{17}} = y^{17-17} = y^0 = 1$

4. a. $(a^4 b)^3 = (a^4)^3 (b)^3 = a^{12} b^3$

   b. $\left(\dfrac{x^3}{y^7}\right)^2 = \dfrac{(x^3)^2}{(y^7)^2} = \dfrac{x^6}{y^{14}}$

5. a. $(-3x^4 y^2)(2xy^5) = -6x^5 y^7$

   b. $(2a^5 b^4)^3 (ab^5)^2 = 2^3 a^{15} b^{12} a^2 b^{10} = 8 a^{17} b^{22}$

   c. $(-x^2)^3 = (-1)^3 x^6 = -x^6$ d. $(-x^3 y^5)(xy^7)(x^4 y) = -x^8 y^{13}$

   e. $(xy^{-5})^{-3} = x^{-3} y^{15} = \dfrac{1}{x^3}(y^{15}) = \dfrac{y^{15}}{x^3}$

f. $-5^2 = -(5)(5) = -25$ g. $(-5)^2 = (-5)(-5) = 25$

h. $-5^{-2} = -\dfrac{1}{5^2} = -\dfrac{1}{25}$ i. $(-5)^{-2} = \dfrac{1}{(-5)^2} = \dfrac{1}{25}$

j. $(7^{-1})^{-1} = 7^{(-1)(-1)} = 7^1 = 7$

k. $-(-11)^0 = -1$ l. $(3^{-4})^3 = 3^{-12} = \dfrac{1}{3^{12}}$

m. $(x^2)^{-3} = x^{-6} = \dfrac{1}{x^6}$

n. $\dfrac{a^{-4}b^3}{a^{-7}b^4} = a^{-4-(-7)}b^{3-4} = a^{-4+7}b^{-1} = a^3 b^{-1} = a^3\left(\dfrac{1}{b}\right) = \dfrac{a^3}{b}$

o. $(a^4 b^{-7})(a^{-3}b^{-2}) = a^{4+(-3)}b^{-7+(-2)} = ab^{-9} = a\left(\dfrac{1}{b^9}\right) = \dfrac{a}{b^9}$

6. -3^2: square 3 and take the opposite of the answer. Result: -9.

   $(-3)^2$: square -3. Result: 9.

7. Again, -3^2 means to first square 3 and then take the opposite of the answer. Result: -9. 3^{-2}, however, means to take the reciprocal of 3^2. So, first we convert to a positive exponent $3^{-2} = \dfrac{1}{3^2}$, then we simplify the 3^2 in the denominator. Result: $\dfrac{1}{9}$.

8. A negative exponent directs us to take the reciprocal of the base and raise the result to the opposite of the original exponent. For example, in 3^{-2} above the reciprocal of the base is $\dfrac{1}{3}$. The opposite of the original exponent would be 2 to get $\left(\dfrac{1}{3}\right)^2 = \dfrac{1}{3^2}$ or $\dfrac{1}{9}$.

9. a. The answers in the table are $32, 16, 8, 4, 2, 1, \dfrac{1}{2}, \dfrac{1}{4}$.

   b. We divide the previous answer by 2 to get the next answer.

   c. The definitions of zero and negative exponents allow the pattern to hold for integer exponents.

Section 5.6 Review

p. 431

1. a. $\dfrac{1}{7}$ b. $-\dfrac{1}{18}$

   c. $-\dfrac{9}{5}$ d. $\dfrac{10}{87} \approx 0.11494\ldots$

   e. $\dfrac{3}{14}$ f. $-\dfrac{1}{8t}$

   g. $\dfrac{1}{2x-3}$ h. $2y+7$

   i. $\dfrac{4x-1}{3x+2}$ j. $\dfrac{-5(Y-2)}{-x+1}$

2. a. $14 \cdot \dfrac{1}{3}$ b. $-8 \cdot \dfrac{5}{2}$

   c. $-5.8 \cdot \dfrac{4}{3}$ d. $\dfrac{4}{5} \cdot -\dfrac{1}{8}$

   e. $-6.4 \cdot \dfrac{2}{7}$ f. $4a \cdot \dfrac{1}{5b}$

   g. $\dfrac{2x}{x-3} \cdot \dfrac{1}{6x}$ h. $\dfrac{3x+12}{4x-1} \cdot \dfrac{x-1}{x+4}$

3. a. $\left(-\dfrac{5}{1}\right)\left(-\dfrac{2}{13}\right) = \dfrac{10}{13}$ b. $\dfrac{29}{3} - \dfrac{19}{5} = \dfrac{145}{15} - \dfrac{57}{15} = \dfrac{88}{15} = 5\dfrac{13}{15}$

   c. $-\dfrac{35}{45} + \left(-\dfrac{12}{45}\right) = -\dfrac{47}{45} = -1\dfrac{2}{45}$ d. $-\dfrac{11}{15}\left(-\dfrac{9}{10}\right) = -\dfrac{11}{5}\left(-\dfrac{3}{10}\right) = \dfrac{33}{50}$

   e. $\dfrac{8}{12x} - \dfrac{15}{12x} = -\dfrac{7}{12x}$ f. $\dfrac{3(y+2)+5(y-3)}{(y-1)(y-3)(y+2)} = \dfrac{8y-9}{(y-1)(y-3)(y+2)}$

   g. $\dfrac{2x}{x-3} \cdot \dfrac{1}{6x} = \dfrac{1}{x-3} \cdot \dfrac{1}{3} = \dfrac{1}{3x-9}$ h. $\dfrac{3(x+4)}{4x-1} \cdot \dfrac{x-1}{x+4} = \dfrac{3}{4x-1} \cdot \dfrac{x-1}{1} = \dfrac{3x-3}{4x-1}$

100

4. a.

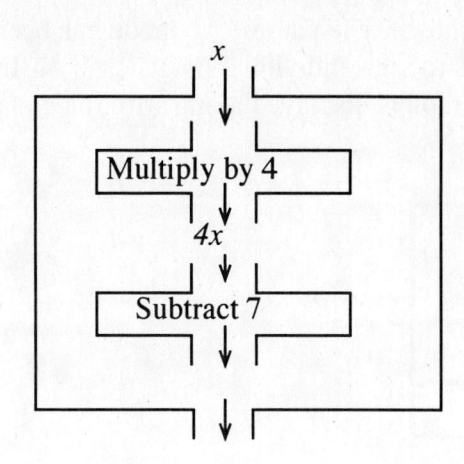

```
        x
        ↓
   ┌─────────────┐
   │ Multiply by 4│
   └─────────────┘
        ↓
        4x
        ↓
   ┌─────────────┐
   │  Subtract 7 │
   └─────────────┘
        ↓
```

b. $y = 4 \cdot \frac{3}{4} - 7 = 3 - 7 = -4$ We are evaluating the function.

c. $4x - 7 = -\frac{2}{3}$

$4x = -\frac{2}{3} + 7 = \frac{19}{3}$ We are solving the first equation shown here.

$x = \frac{1}{4} \cdot \frac{19}{3} = \frac{19}{12} = 1\frac{7}{12}$

d. $4x - 7 = 5.47$
$4x = 12.47$ Solving an equation again.
$x = \frac{12.47}{4} = 3.1175$

e. $y = 4\left(-2\frac{6}{7}\right) - 7 = 4\left(-\frac{20}{7}\right) - 7 = -\frac{80}{7} - \frac{49}{7} = -\frac{129}{7} = -18\frac{3}{7}$ Evaluating.

5. a. **Original and Sale Price**

Original Price (P)	Sale Price (S)
10	6
15	9
20	12
25	15
30	18

b. $S(P) = 0.6P$

c. The graph as produced on the TI-83 is shown. The parametric mode has been chosen to display the graph because it allows us to view individual points easily using the trace feature. One of the points on the graph is displayed along with the equations entered into the Y= menu.

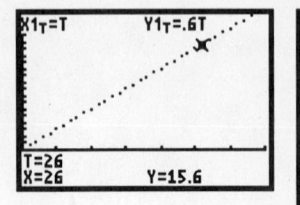

6. Let d represent distance in miles and t represent time in hours.

   a. 10–12: $\dfrac{\Delta d}{\Delta t} = \dfrac{78334 - 78239}{12 - 10} = \dfrac{95}{2} = 47.5$ m.p.h.

   12–1:15: $\dfrac{\Delta d}{\Delta t} = \dfrac{78334 - 78334}{1\frac{1}{4}} = 0$ m.p.h.

   1:15–2:30: $\dfrac{\Delta d}{\Delta t} = \dfrac{78383 - 78334}{1\frac{1}{4}} = \dfrac{49}{\frac{5}{4}} = 49\left(\dfrac{4}{5}\right) = 39.2$ m.p.h.

   2:35–5:10: $\dfrac{\Delta d}{\Delta t} = \dfrac{78489 - 78383}{2\frac{7}{12}} = \dfrac{106}{\frac{29}{12}} = 106\left(\dfrac{12}{29}\right) \approx 44$ m.p.h.

   b. $\dfrac{\Delta d}{\Delta t} = \dfrac{78489 - 78239}{7\frac{1}{6}} = \dfrac{250}{\frac{43}{6}} = 50\left(\dfrac{6}{43}\right) \approx 34.9$ m.p.h. The average speed over the entire trip was about 35 m.p.h.

7. a. slope $= \dfrac{4 - (-1)}{2 - 5} = \dfrac{5}{-3} = -\dfrac{5}{3}$

   b. slope $= \dfrac{-6 - (-9)}{11 - (-4)} = \dfrac{3}{15} = \dfrac{1}{5}$

   c. Since the change in the first coordinates is zero the line determined by these two points is not defined. The line is vertical. In general, vertical lines do not have a slope.

8. a. −1   b. 9

c. $\dfrac{2}{3}$

d. $-\dfrac{6}{5}$

9. a. k^{99}
   b. 1
   c. -1
   d. $\dfrac{1}{p^{23}}$
   e. a^7
   f. $\dfrac{1}{b^{16}}$
   g. Can't be simplified because the bases are different.
   h. $\dfrac{1}{(x^4y^7)^2} = \dfrac{1}{x^8y^{14}}$
   i. $\left(\dfrac{b^3}{a}\right)^4 = \dfrac{b^{12}}{a^4}$

10. a. $(-a^{20}b^{35})(a^{14}b^6) = -a^{34}b^{41}$
    b. $\dfrac{1}{x^7y^5}$
    c. x^2y^3
    d. $(5^{-2}x^2y^{-6})(3^4x^{10}y^2) = \dfrac{81x^{12}}{25y^4}$
    e. $(x^2y^3)^3 = x^6y^9$
    f. $\dfrac{1}{x+y}$
    g. $\dfrac{1}{x} + \dfrac{1}{y}$

11. You could compare the strength of each pot of coffee by doing a multiplicative comparison. Compare the ratio of the number of spoons of coffee to the amount of water in each of the pots. The one with the largest coffee to water ratio is the strongest.

12. The ratios to consider are $\dfrac{1.19}{12}$ or about $0.099 per ounce and $\dfrac{1.49}{16}$ which is about $0.093 per ounce. The soup in the 16 ounce can is cheaper per ounce. Other factors which might influence your decision are the amount of soup in the can, you may not want 16 ounces, or the kind of soup.

13. The gas mileage of the basic car is $\dfrac{275}{12}$ or $22\dfrac{11}{12}$ miles per gallon. The gas mileage for the deluxe version is $\dfrac{371}{18}$ or $20\dfrac{11}{18}$ miles per gallon. As expected the basic version gets better mileage.

103

CHAPTER 6

Real Numbers: Completing a Mathematical System

Section 6.1 Real Numbers and the Algebraic Extension

p. 446

2. a. The whole numbers are 3, 5, and 0.

   b. The integers are 3, −7, 5, −$\sqrt{9}$, −$\frac{15}{3}$, 0.

   c. The rational numbers are 3, −7, 5, −$\sqrt{9}$, −$\frac{15}{3}$, 0, −$\frac{2}{3}$, $\frac{3}{5}$, $\frac{13}{7}$, −$\frac{6}{11}$, $\sqrt{\frac{9}{4}}$, $\sqrt{0.25}$, 2^{-3}, 1.3^2,

   |2.3|, |−3.1|, −|4.6|, −|−3.7|.

   d. The irrational numbers are $\sqrt{7}$, −$\sqrt{2}$, $\sqrt{\frac{2}{3}}$, $\sqrt{5.72}$.

   e. All the numbers in the collection are real numbers.

3. a. A: 3 = 3; B: |2.3| = 2.3; C: −7 = −7; D: −$\frac{2}{3}$ ≈ −0.6667; E: $\frac{3}{5}$ = 0.6

   F: $\sqrt{7}$ ≈ 2.6458; G: |−3.1| = 3.1; H: 5 = 5; I: −$\sqrt{2}$ ≈ −1.414; J: $\frac{13}{7}$ ≈ 1.8571;

   K: −$\frac{6}{11}$ ≈ −0.5454; L: $\sqrt{\frac{2}{3}}$ ≈ 0.8165; M: −$\sqrt{9}$ = −3; N: $\sqrt{5.72}$ ≈ 2.3917;

   O: −|4.6| = −4.6; P: −$\frac{15}{3}$ = −5; Q: 0 = 0; R: $\sqrt{\frac{9}{4}}$ = 1.5; S: $\sqrt{0.25}$ = 0.5;

   T: 2^{-3} = 0.125; U: 1.3^2 = 1.69; V: −|−3.7| = −3.7.

4.

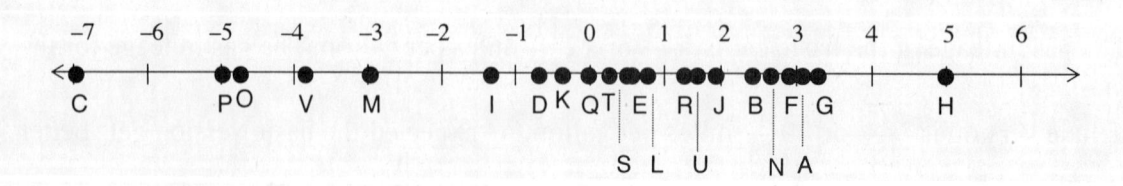

5. C, P, O, Y, M, I, D, K, Q, T, S, E, L, R, U, J, B, N, F, A, G, H. Note: These letters are the labels used in numbers 3 and 4, above.

6. a. The whole numbers are closed with respect to addition and multiplication.

   b. The integers are closed with respect to addition, subtraction, multiplication, and oppositing.

   c. The rational numbers are closed with respect to addition, subtraction, multiplication, division, oppositing, and reciprocaling.

   d. The real numbers are closed with respect to addition, subtraction, multiplication, division, oppositing and reciprocaling.

7. For all parts, all numbers displayed are rational numbers. There are an infinite number of real numbers between any two real numbers though the calculator can only display a finite number of them.

8. Some common numbers used to estimate π are 3.14 and $\frac{22}{7}$.

   a. Comparing the approximations to the first two hundred digits of π printed in the section, the rational number $\frac{355}{113}$ is the best approximation of the ones mentioned. Both the approximation 3.14 and $\frac{22}{7} \approx 3.142857$ are correct to two decimal places. But $\frac{355}{113} \approx 3.14159292$ is correct to six decimal places.

   b. The more decimal places in the approximation that match the actual decimal value, the better the approximation.

   c. The number π is an irrational number. Its decimal representation is non–terminating, non–repeating. Thus we can never write the complete decimal value. The only way to represent π exactly is to write π.

   d. Approximations of π are usually rational as we usually use a fraction or a terminating decimal to approximate π. These are rational numbers.

9. a. $p = -\frac{11}{5}$ Rational numbers. b. $t = 0$ Whole numbers.

   c. $y = 1$ Whole numbers. d. $x = 1$ Whole numbers.

10. a. $(2x^{-5})(3^2 x^2) = 18x^{-3} = \frac{18}{x^3}$

b. $(4x^{-1}y^4)^3(3xy^{-4})^{-2} = (4^3)x^{-3}y^{12}3^{-2}x^{-2}y^8 = \dfrac{64y^{20}}{9x^5}$

c. $\dfrac{3y^2}{4x^5}$ d. $\dfrac{a^2}{2b}$

Section 6.2 The Square Root Function

p. 457

2. a. $\sqrt{7} = 2.6457513$ b. $-\sqrt{12} = -3.4641016$

c. $\sqrt{81} = 9$ d. $-\sqrt{49} = -7$

e. $\sqrt{-49}$ undefined   f. $2\sqrt{11} = 6.6332496$

g. $-3\sqrt{5} = -6.7082039$ h. $4\sqrt{9} = 12$

i. $3\sqrt{-2}$ undefined   j. $-7\sqrt{121} = -77$

3.   Function machine for $3\sqrt{-2}$.

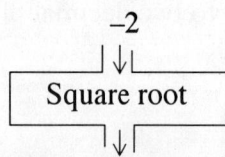

-2
↓
| Square root |
↓

No output since –2 is not in the domain of the function.

Function machine for $-7\sqrt{121}$.

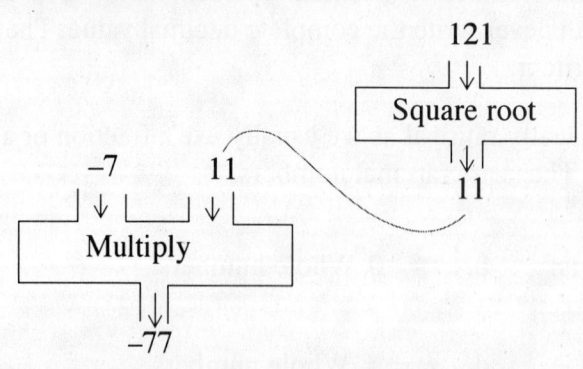

121
↓
| Square root |
↓
11

-7 11
↓   ↓
| Multiply |
↓
-77

4. a. **Table 2: Absolute Value Versus the Square Root of a Square**

x	$abs(x)$	$\sqrt{x^2}$
−7	7	7
−5	5	5
−3	3	3
−1	1	1
0	0	0
2	2	2
4	4	4
6	6	6

b. and c.

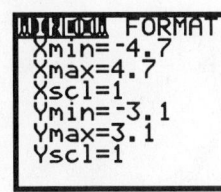

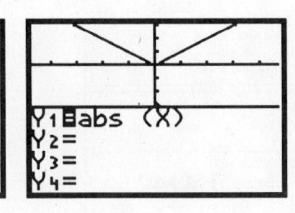

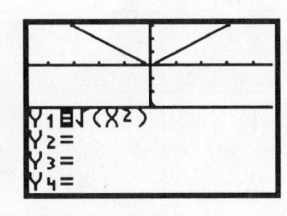

```
WINDOW FORMAT
Xmin=-4.7
Xmax=4.7
Xscl=1
Ymin=-3.1
Ymax=3.1
Yscl=1
```

```
Y1=abs (X)
Y2=
Y3=
Y4=
```

```
Y1=√(X²)
Y2=
Y3=
Y4=
```

d. The function $abs(x)$ and the function $\sqrt{x^2}$ are the same function.

5. The absolute value function.

6. a. **Table 3: Rate of Change of the Squaring Function**

x	x^2	$\Delta(x^2)$
1	1	3
2	4	5
3	9	7
4	16	9
5	25	

b. **Table 4: Rate of Change of the Square Root Function**

x	$\sqrt{x}$	$\Delta(\sqrt{x})$
1	1	.41421
2	1.4142	.31784
3	1.7321	.26795
4	2	.23607
5	2.23607	

c. The outputs are both increasing since the finite differences are positive.

d. The finite differences in part a are increasing at a constant rate meaning the outputs are increasing faster and faster. The finite differences in part b are decreasing meaning the outputs are increasing a slower and slower.

7. a. **Comparing square roots with raising a number to the one-half power**

x	$\sqrt{x}$	$x^{1/2}$
1	1	1
2	1.4142	1.4142
3	1.7321	1.7321
4	2	2
5	2.23607	2.23607
6	2.44949	2.44949
7	2.64575	2.64575
8	2.828427	2.828427

b. and c.

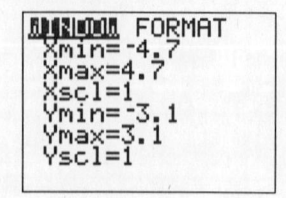

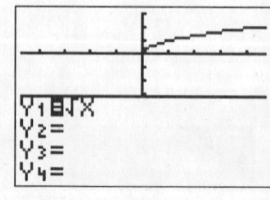

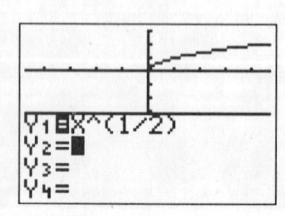

d. They are the same.

e. $(x^{1/2})^2 = (x^{1/2})(x^{1/2}) = x^{\frac{1}{2}+\frac{1}{2}} = x^1 = x$. So $x^{1/2}$ is the square root of x.

So $x^{1/2} = \sqrt{x}$.

8. a. The answers are getting closer to 1.

   b. These also approach 1 as the number of repetitions is increased.

9. a. $3^{1/2} \approx 1.73205$ b. $16^{1/2} = 4$

   c. $100^{1/2} = 10$ d. $17^{1/2} \approx 4.12311$

   e. $-5^{1/2} \approx -2.23607$ f. $-49^{1/2} = -7$

   g. $(-4)^{1/2}$ undefined

10. a. $(3, 8)$ and $(5, 2)$. $\Delta x = 5 - 3 = 2, \Delta y = 2 - 8 = -6$

   $distance = \sqrt{(2)^2 + (-6)^2} = \sqrt{4 + 36} = \sqrt{40} \approx 6.32455$

   b. $(-6, 3)$ and $(4, -9)$. $\Delta x = 4 - (-6) = 10, \Delta y = -9 - 3 = -12$

   $distance = \sqrt{(10)^2 + (-12)^2} = \sqrt{100 + 144} = \sqrt{244} \approx 15.6205$

   c. $(-3, -5)$ and $(-7, 2)$. $\Delta x = -7 - (-3) = -4, \Delta y = 2 - (-5) = 7$

   $distance = \sqrt{(-4)^2 + (7)^2} = \sqrt{16 + 49} = \sqrt{65} \approx 8.06226$

   d. $(8, -5)$ and $(2, -5)$. $\Delta x = 2 - 8 = -6, \Delta y = -5 - (-5) = 0$

   $distance = \sqrt{(-6)^2 + (0)^2} = \sqrt{36 + 0} = \sqrt{36} = 6$

   e. $(4, 3)$ and $(4, -6)$. $\Delta x = 4 - (4) = 0, \Delta y = -6 - 3 = -9$

   $distance = \sqrt{(0)^2 + (-9)^2} = \sqrt{0 + 81} = \sqrt{81} = 9$

11. a. Let $d = 30$, then $t(30) = \sqrt{\dfrac{30}{16}} \approx 1.4$. It will take about 1.4 seconds to fall 30 feet.

   b. One mile equals 5280 feet. So $d = 5280$. Then $t(5280) = \sqrt{\dfrac{5280}{16}} \approx 18$. It will take about 18 seconds to fall one mile.

   c. distance fallen = $9500 - 3000 = 6500$. So $d = 6500$. Then $t(6500) = \sqrt{\dfrac{6500}{16}} \approx 20$. The skydiver will free fall for about 20 seconds.

12. $S(d) = \sqrt{30fd}$. In this problem, $f = 0.83$. So $S(d) = \sqrt{30(0.83)d}$.

   a. Let $d = 100$. Then $S(100) = \sqrt{30(0.83)(100)} \approx 50$. The minimum speed of the car was about 50 m.p.h. if the skid marks are 100 feet long.

   b. Let $d = 60$. Then $S(60) = \sqrt{30(0.83)(60)} \approx 38.7$. The drive was going about 39 m.p.h. so a ticket is in order.

   c. Use trial and error to find d if $S(d) = 60 \Rightarrow \sqrt{30(0.83)d} = 60$. The length of the skid marks is about 145 feet.

13. a. $9y^2 - 49z^2$ b. $9y^2 + 42yz + 49z^2$

   c. $-14z$ d. $x^3 + 1$

14. a. $\dfrac{6a}{3(a+3b)} = \dfrac{2a}{a+3b}$ b. $\dfrac{2(2x-4)}{2(3x-6)} = \dfrac{2x-4}{3x-6}$

   c. $\dfrac{x+3}{7(x-3)}$ d. This one is reduced.

Section 6.3 Classes of Basic Functions

p. 480

2. a. Linear, quadratic, absolute value, and square root.

   b. Opposing and reciprocal.

   c. Quadratic.   d. Square root.

e. Oppositing.   f. Reciprocal.

3. a. $x = 17$.   b. $x = -4, x = 4$.

    c. $x = -13, x = 13$.   d. $x = -13$

    e. $x = \dfrac{17}{2}$ f. $x = 9$

4. Constant function: The tuition problem in which students were charged a flat rate for all the credit hours over 12 taken in a semester.

    Linear function: Okimbe's Bulls T–shirts.

    Quadratic: The travels of the falling corned beef sandwich.

    Opposite: Calculator golf.

    Absolute value: The stock market problem.

    Reciprocal: The 240–mile trip.

    Square root: The area of the dog pen.

5. The algebraic representation for the Okimbie profit function is $y = 10x - 1000$. See exploration number 4 on page 307 in Section 4.4.

    a. The vertical intercept is $(0, -1000)$. This means that if Okimbie sells no shirts he is in the hole $1000, the cost of the shirts.

    b. The horizontal intercept is found by solving the equation $10x - 1000 = 0$ to get $x = 100$. This is the number of shirts Okimbie must sell to just break even.

6. a. **Table 12: Input-Output Table**

x	−5	−4	−3	−2	−1	0	1	2	3	4	5
$y(x)$	−6	−4	−2	0	2	4	6	8	10	12	14

    b. Vertical intercept: 4   Horizontal intercept: −2.

    c. The graph of $y(x) = 2x + 4$ appears in the standard viewing window.

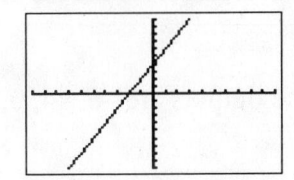

111

d. Increasing since, as the input increases, the output also increases. This can be determined by reading the table of values from left to right.

7. a. **Table 13 Input-Output Table**

x	−5	−4	−3	−2	−1	0	1	2	3	4	5
$y(x)$	16	14	12	10	8	6	4	2	0	−2	−4

b. Vertical intercept: 6   Horizontal intercept: 3.

c. The graph of $y(x) = -2x + 6$ appears in the standard viewing window.

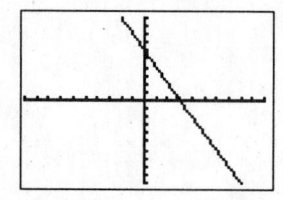

d. Decreasing since, as the input increases, the output decreases. This can be determined by reading the table of values from left to right.

e. The graph is a line that intersects the horizontal axis at three and the vertical axis at six.

8. a. **Table 14: Input-Output Table**

x	−5	−4	−3	−2	−1	0	1	2	3	4	5
$y(x)$	14	6	0	−4	−6	−6	−4	0	6	14	24

b. Vertical intercept: −6   Horizontal intercepts: −3 and 2.

c. The graph of $y(x) = x^2 + x - 6$ appears in the standard viewing window.

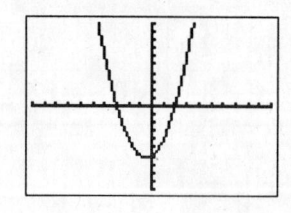

d. Neither. As the input increases up to −1, the output decreases. As the input increases from zero, the output also increases.

e. The inputs are −5, −4, −3, −2, −1, 0, 1, 2, 3, 4, 5. The outputs are −6, −4, 0, 6, 14, 24.

f. The domain is all real numbers.

g. The range is all real numbers greater than or equal to approximately –6.

9. The functions in Explorations 5 and 6 have one horizontal intercept while the function in Exploration 7 has two. This occurs since the functions given in Explorations 5 and 6 are linear functions while the function given in Exploration 7 is a quadratic function.

10. As x increases, $\frac{1}{x}$ decreases to 0. This is true since we are increasing the size of the denominator making the value of the fraction smaller and smaller.

11. a. In y stages the spaceship will travel $11y$.

    b. The function $Distance(y) = 11y$ is a linear function.

    c. The new distance function is $Earth(y) = 11y + 4$.

Section 6.4 Linear Functions

p. 504

2. a. slope = 3　　　　　　　　　　　b. vertical intercept: $(0, -12)$

    c. horizontal intercept: $(4, 0)$　　　d. The graph is increasing.

    e. $3x - 12 = 7$
        $X = \frac{19}{3}$

3. a. slope = –5　　　　　　　　　　　b. vertical intercept: $(0, 10)$

    c. horizontal intercept: $(2, 0)$　　　d. The graph is decreasing.

    e. $-5x + 10 = 7$
        $X = \frac{3}{5}$

4. a. slope = 0　　　　　　　　　　　b. vertical intercept: $(0, -3)$

    c. There is no horizontal intercept.　d. The graph is neither increasing nor decreasing.

    e. The output cannot be 7 for this function.

5. a. slope = –2　　　　　　　　　　　b. vertical intercept: $(0, -5)$

c. horizontal intercept: $\left(-\dfrac{5}{2}, 0\right)$ d. The graph is decreasing.

e. $-2x - 5 = 7$
    $x = -6$

6. a. slope $= \dfrac{2}{3}$ b. vertical intercept: $(0, -5)$

    c. horizontal intercept: $\left(\dfrac{15}{2}, 0\right)$ d. The graph is increasing.

    e. $\dfrac{2}{3}x - 5 = 7$
        $x = 18$

7. a. slope $= 7$ b. vertical intercept: $\left(0, \dfrac{8}{5}\right)$

    c. horizontal intercept: $\left(-\dfrac{8}{35}, 0\right)$ d. The graph is increasing.

    $7p + \dfrac{8}{5} = 7$

    e. $7p = 7 - \dfrac{8}{5} = \dfrac{35}{5} - \dfrac{8}{5} = \dfrac{27}{5}$

    $p = \dfrac{27}{5} \cdot \dfrac{1}{7} = \dfrac{27}{35}$

8. a. The graph of $y(x) = 3x + 4$ appears in the standard viewing window.

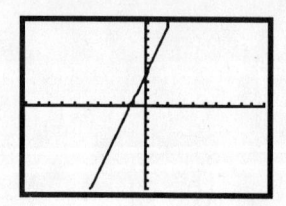

b.

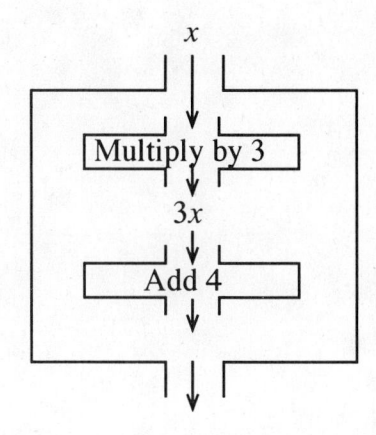

[diagram: x → Multiply by 3 → 3x → Add 4 →]

c. $y(x) = 3x + 4$ d. horizontal intercept: $\left(-\frac{4}{3}, 0\right)$

e. $y(x) = 3x + 6$ f. $y(x) = -\frac{1}{3}x + 4$

9. a. The graph of $y(x) = -2x + 3$ appears in the standard viewing window.

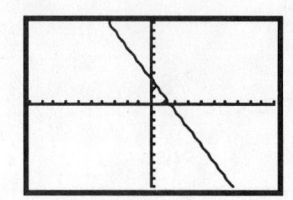

b.

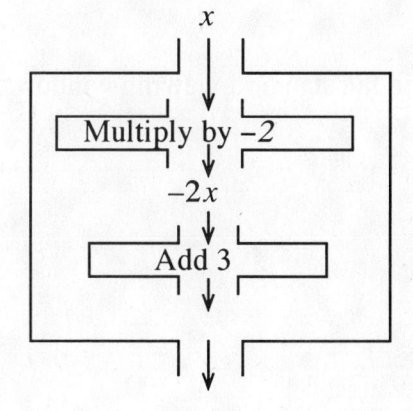

[diagram: x → Multiply by −2 → −2x → Add 3 →]

c. $y(x) = -2x + 3$ d. horizontal intercept: $\left(\frac{3}{2}, 0\right)$

e. $y(x) = -2x + 5$ f. $y(x) = \frac{1}{2}x + 3$

10. a. The graph of $y(x) = \frac{3}{2}x - 4$ appears in the standard viewing window.

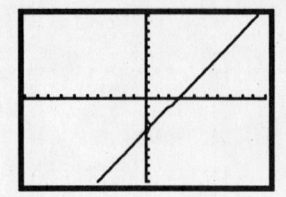

b.

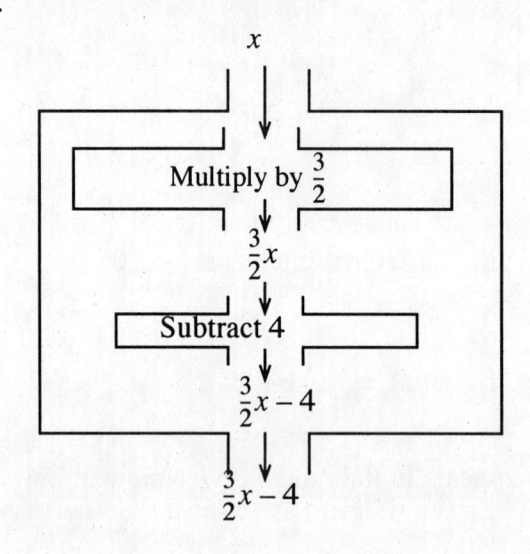

c. $y(x) = \frac{3}{2}x - 4$

d. horizontal intercept: $\left(\frac{8}{3}, 0\right)$

e. $y(x) = \frac{3}{2}x - 2$

f. $y(x) = -\frac{2}{3}x - 4$

11. a. The graph of $y(x) = -\frac{3}{5}x - 1$ appears in the standard viewing window.

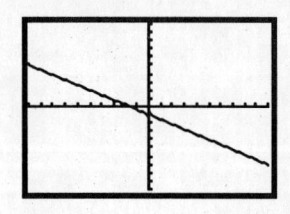

b.

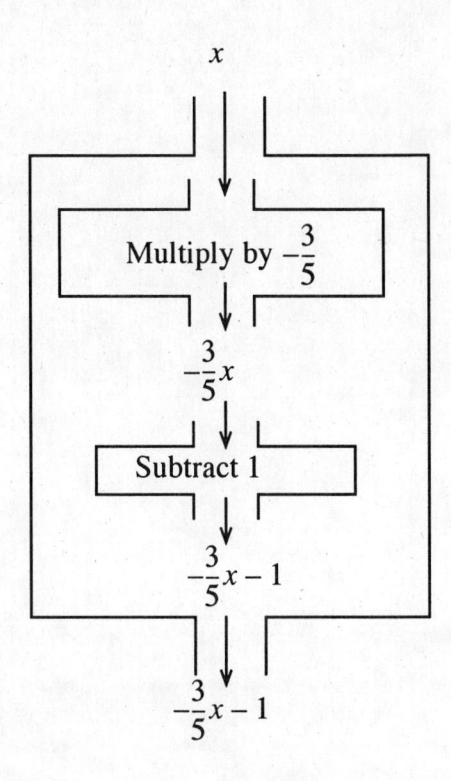

```
         x
         ↓
   ┌───────────┐
   │Multiply by −3/5│
   └───────────┘
         ↓
        −3/5 x
         ↓
   ┌───────────┐
   │ Subtract 1 │
   └───────────┘
         ↓
       −3/5 x − 1
         ↓
       −3/5 x − 1
```

c. $y(x) = -\dfrac{3}{5}x - 1$ d. horizontal intercept: $\left(-\dfrac{5}{3}, 0\right)$

e. $y(x) = -\dfrac{3}{5}x + 1$ f. $y(x) = \dfrac{5}{3}x - 1$

12. a. The graph of $y(x) = \dfrac{7}{4}x - \dfrac{3}{2}$ appears in the standard viewing window.

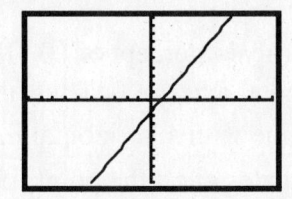

117

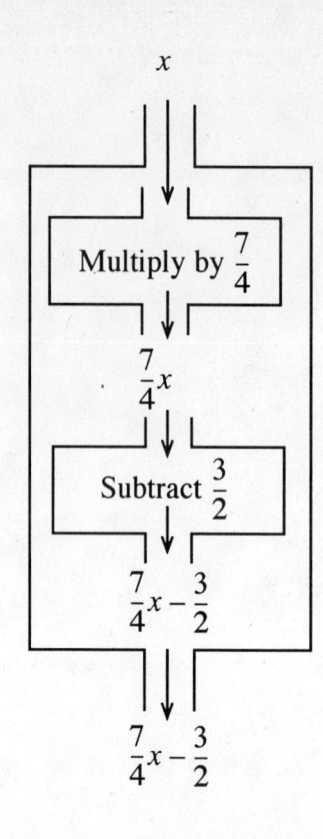

b.

```
         x
         ↓
   ┌───────────┐
   │Multiply by 7/4│
   └───────────┘
         ↓
        7/4 x
         ↓
   ┌───────────┐
   │Subtract 3/2│
   └───────────┘
         ↓
     7/4 x − 3/2
         ↓
     7/4 x − 3/2
```

c. $y(x) = \dfrac{7}{4}x - \dfrac{3}{2}$ d. horizontal intercept: $\left(\dfrac{6}{7}, 0\right)$

e. $y(x) = \dfrac{7}{4}x + \dfrac{1}{2}$ f. $y(x) = -\dfrac{4}{7}x - \dfrac{3}{2}$

13. a. $y(x) = 0$ b. $x = 0$

c. The horizontal axis is a horizontal line. It has a vertical intercept of $(0, 0)$. In addition all of the y-coordinates of the points on the horizontal axis are 0. Thus, the equation is $y = 0$. The vertical axis is a vertical line with a horizontal intercept of $(0, 0)$. One should also observe that the x-coordinates of all the points on the vertical axis are 0. The equation is $x = 0$.

14. a. $m = \dfrac{\Delta y}{\Delta x} = \dfrac{4-7}{2-3} = \dfrac{-3}{-1} = 3$ b. $m = \dfrac{\Delta y}{\Delta x} = \dfrac{-3-6}{5-(-2)} = \dfrac{-9}{7} = -\dfrac{9}{7}$

c. $m = \dfrac{\Delta y}{\Delta x} = \dfrac{6-2}{9-4} = \dfrac{4}{5}$

15. The output at the horizontal intercept for the cookie problem is 120 cookies. This represents the breakeven point.

16. a. Vertical intercept is (0, –10). b. Slope is approximately 5.

   c.

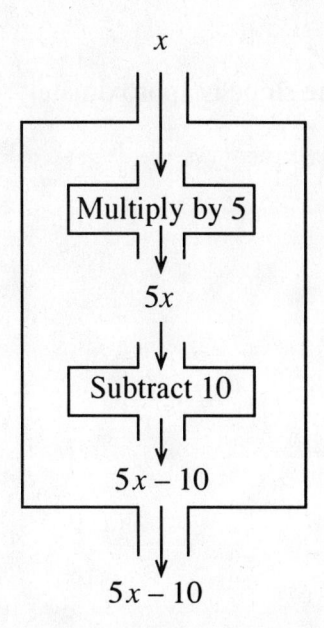

   ```
   x
   ↓
   [Multiply by 5]
   ↓
   5x
   ↓
   [Subtract 10]
   ↓
   5x – 10
   ↓
   5x – 10
   ```

   d. Approximate equation is $y(x) = 5x - 10$.

   e. The horizontal intercept is (2, 0).

17. a. Vertical intercept is (0, 100). b. Slope is approximately –25.

   c.

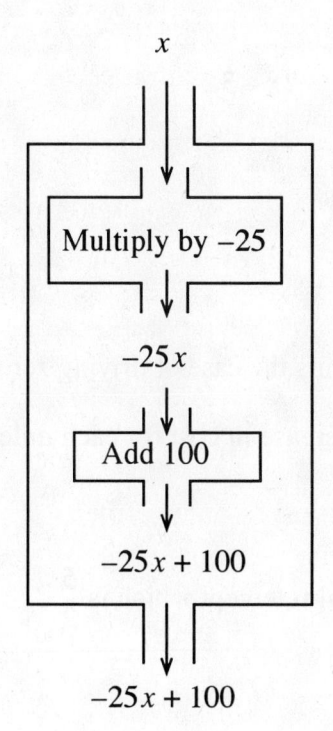

   ```
   x
   ↓
   [Multiply by –25]
   ↓
   –25x
   ↓
   [Add 100]
   ↓
   –25x + 100
   ↓
   –25x + 100
   ```

d. The approximate equation is $y(x) = -25x + 100$.

e. The horizontal intercept is $(4, 0)$.

18. a. The vertical intercept is –3.   b. The slope is approximately –2.

    c.

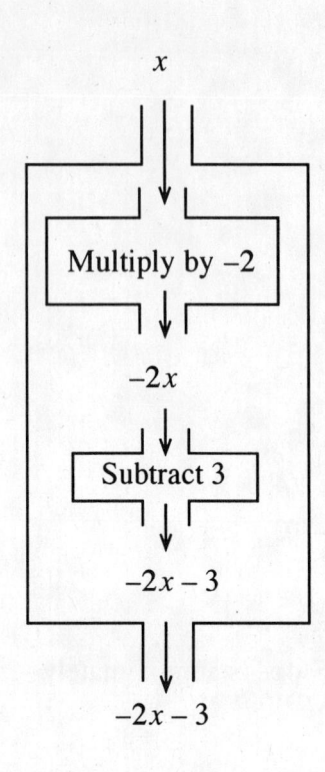

    ```
         x
         ↓
    ┌─────────────┐
    │ Multiply by –2 │
    └─────────────┘
         ↓
        –2x
         ↓
    ┌─────────┐
    │ Subtract 3 │
    └─────────┘
         ↓
       –2x – 3
         ↓
       –2x – 3
    ```

    d. The approximate equation is $y(x) = -2x - 3$.

    e. The horizontal intercept is $(-1.5, 0)$.

19. a. $T(m) = 0.20m + 15$

    b. The vertical intercept is 15. It represents the case of driving zero miles.

    c. The slope is 0.20. It represents the increase in cost for each mile driven.

20. a. $y(x) = 0$ means we want the vertical intercept which is $\frac{5}{2}$.

    b. Solving $2x - 5 = 3$ we get $x = 4$.

    c. Solving $2x - 5 = 13$ gives us $x = 9$.

21. a. $P(c) = 10 \Rightarrow 0.25c - 30 = 10 \Rightarrow c = 160$. So we must sell 160 cookies to realize a profit of $10.

    b. $P(c) = 100 \Rightarrow 0.25c - 30 = 100 \Rightarrow c = 520$. So we must sell 520 cookies to realize a profit of $100.

22. a–d. Yes to both.

    e. The domain is all real numbers.

    f. The range is all real numbers.

Section 6.5 Quadratic Functions

p. 521

2. a. Narrower.
   b. Shifted up 4 units.
   c. Reflected about the x–axis and narrower.
   d. Shifted down 7 units.
   e. Shifted up one unit and wider.
   f. Shifted down 4 units and narrower.
   g. Reflected about the x–axis and shifted down 3 units.

3. $x - (-6)$ can be written as an addition by changing the operation to addition and finding the opposite of –6. Doing both results in $x + 6$. Therefore, $x - (-6)$ and $x + 6$ are equivalent.

4. a. Zeros: 6, –6. x–value of vertex: 0.

   b. Zeros: 9, –7. x–value of vertex: 1, found by computing $\frac{9 + (-7)}{2} = \frac{2}{2} = 1$.

   c. Zeros: –3, –5. x–value of vertex: –4.

   d. Zeros: 11, 7. x–value of vertex: 9.

   e. Zeros: –5, 8. x–value of vertex: 1.5.

   f. Zeros: 2. x–value of vertex: 2.

5. a. $y(x) = (x - 1)(x - 4)$ So $y(x) = x^2 - 5x + 4$

b. $y(x) = (x+3)(x+2)$ So $y(x) = x^2 + 5x + 6$

c. $y(x) = (x-4)(x+6)$ So $y(x) = x^2 + 2x - 24$

d. $y(x) = (x+5)(x-7)$ So $y(x) = x^2 - 2x - 35$

e. $y(x) = (x+5)(x-5)$ So $y(x) = x^2 - 25$

f. $y(x) = (x-7)(x-7)$ So $y(x) = x^2 - 4x + 49$

6. a. Zeros: –4, 4

   $y(x) = (x+4)(x-4)$. So $y(x) = x^2 - 16$.

   b. Zeros: 1, 6.

   $y(x) = (x-1)(x-6)$. So $y(x) = x^2 - 7x + 6$.

   c. Zeros: –5, –2.

   $y(x) = (x+5)(x+2)$. So $y(x) = x^2 + 7x + 10$.

   d. The graph is a parabola that opens upward. The vertical intercept is at ten. The horizontal intercepts are at –2 and –5. The vertex is at approximately (–3.5, –5).

7. The perimeter would be given by the value of $2n$.

8. a. $x^2 + 3x + 3x + 9 = x^2 + 6x + 9$

   b. $x + 3$ See part f, below for the diagram.

   c. $x^2 + 3x + 3x + 9 = x^2 + 6x + 9$

   d. They are the same.

   e. $x^2 + 3x + 2x + 6$ It is not a square since the lengths of the sides are not all the same.

	x	3
x	x^2	$3x$
2	$2x$	6

122

f.

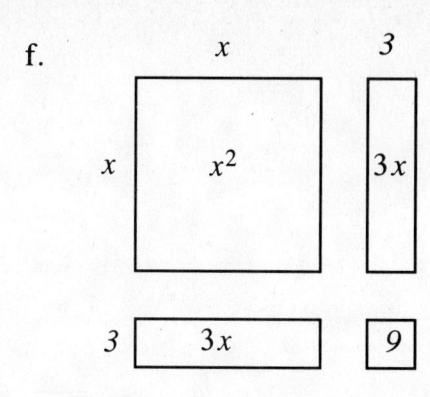

	x	3
x	x^2	$3x$
3	$3x$	9

They do form a square when put together. The length of the sides for the square would be $x + 3$.

9. a. $4x - 7 = -9$

$$x = -\frac{1}{2}$$

b. $5 - 3x = \frac{2}{3}$

$$-3x = \frac{2}{3} - 5 = -\frac{13}{3}$$

$$x = \left(-\frac{13}{3}\right)\left(-\frac{1}{3}\right) = \frac{13}{9}$$

c. $\frac{2}{5}t + 3 = \frac{5}{7}$

$$\frac{2}{5}t = \frac{5}{7} - 3 = -\frac{16}{7}$$

$$t = -\frac{16}{7} \cdot \frac{5}{2} = -\frac{8}{7} \cdot \frac{5}{1} = -\frac{40}{7}$$

d. $\frac{7}{6} - 2y = -5$

$$-2y = -5 - \frac{7}{6} = -\frac{37}{6}$$

$$y = -\frac{37}{6} \cdot \left(-\frac{1}{2}\right) = \frac{37}{12}$$

10. a. Slope = 2   Vertical Intercept: (0, 7)   Horizontal Intercept: $\left(-\frac{7}{2}, 0\right)$

b. Slope = −9   Vertical Intercept: (0, 5)   Horizontal Intercept: $\left(\frac{5}{9}, 0\right)$

c. Slope = $\frac{3}{4}$ Vertical Intercept: $\left(0, -\frac{2}{5}\right)$ Horizontal Intercept: $\left(\frac{8}{15}, 0\right)$

d. Slope = 5   Vertical Intercept: (0, −11)   Horizontal Intercept: $\left(\frac{11}{5}, 0\right)$

Section 6.6 Review

p. 526

1. a. The whole numbers are 0 and 18.

   b. Integers: $0, 18, -9, -\sqrt{4}, -\dfrac{24}{6}$

   c. Rational Numbers: $0, 18, -9, -\sqrt{4}, -\dfrac{24}{6}, -\dfrac{4}{5}, \dfrac{24}{9}, \dfrac{3}{5}, |-5.4|, 5^{-2}, \sqrt{\dfrac{4}{9}}, 0.8^2, \sqrt{0.49}$

   d. Irrational Numbers: $\sqrt{8}, -\sqrt{5}, \sqrt{\dfrac{5}{6}}, \sqrt{1.04}$

   e. All are real numbers.

2 & 3. The labels are for showing the order requested in exploration 3 and for reference in exploration 4, appearing below.

A: -9 B: $-\dfrac{24}{6} = -4$ C: $-\sqrt{5} \approx -2.236068$ D: $-\sqrt{4} = -2$

E: $-\dfrac{4}{5} = -0.8$ F: $-\dfrac{1}{12} = -0.08333...$ G: $0 = 0$ H: $5^{-2} = \dfrac{1}{25} = 0.04$

I: $\dfrac{3}{5} = 0.6$ J: $0.8^2 = 0.64$ K: $\sqrt{\dfrac{4}{9}} = \dfrac{2}{3} = 0.666...$

L: $\sqrt{0.49} = 0.7$ M: $\sqrt{\dfrac{5}{6}} \approx 0.912871$ N: $\sqrt{1.04} \approx 1.01983903$

O: $\dfrac{24}{9} = 2\dfrac{2}{3} = 2.666...$ P: $\sqrt{8} \approx 2.8284271$ Q: $|-5.4| = 5.4$ R: $18 = 18$

4. The letters used to label the numbers above are used to indicate their position in the graph below.

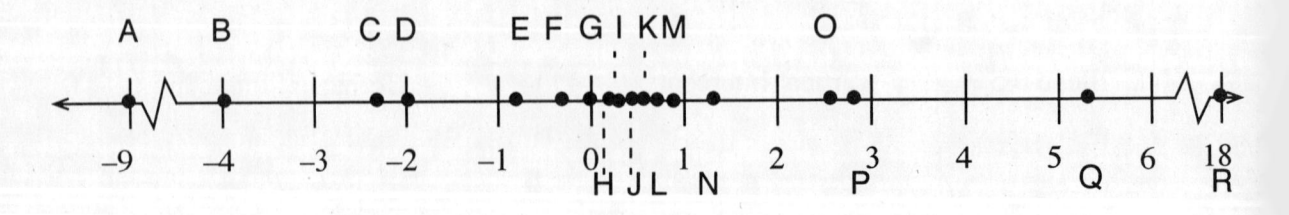

The position of the values on the number line is approximate.

124

5. The only one that is not a real number is g because the square of any real number must be greater than or equal to zero. So, there is no real number we could square and get –64.

   a. $\sqrt{3} \approx 1.732$
   b. $5\sqrt{17} \approx 20.616$
   c. $\sqrt{36} = 6$
   d. $-\sqrt{64} = -8$
   e. $-2\sqrt{6} \approx -4.899$
   f. $-\sqrt{22} \approx -4.690$
   g. not a real number
   h. $7\sqrt{4} = 14$

   6 represents the only one which is not a real number because there is no real number we could square to get –64.

6. a. $3^{1/2} \approx 1.732$
   b. $-64^{1/2} = -8$
   c. $19^{1/2} \approx 4.359$
   d. $144^{1/2} = 12$
   e. not a real number
   f. $64^{1/2} = 8$
   g. $-6^{1/2} \approx 2.449$

7. We use the formula $\sqrt{(\Delta x)^2 + (\Delta y)^2}$ to find the length of each line segment.

   a. $\sqrt{(-6-9)^2 + (7-5)^2} = \sqrt{(-15)^2 + 2^2} = \sqrt{225+4} = \sqrt{229} \approx 15.133$

   b. $\sqrt{(4-(-7))^2 + (-8-(-15))^2} = \sqrt{11^2 + 7^2} = \sqrt{121+49} = \sqrt{170} \approx 13.038$

8. a. $x = 52$
   b. $x^2 = 5$
      $x = \pm\sqrt{5}$
   c. $|x| = 4$
      $x = \pm 4$
   d. $-x = -(9)$
      $x = 9$
   e. $\dfrac{1}{x} = \dfrac{13}{8}$
      $x = \dfrac{8}{13}$
   f. $\sqrt{x} = 4$
      $x = 16$

9. a. **Table 1: Input-Output Table**

x	−5	−4	−3	−2	−1	0	1	2	3	4	5
$y(x)$	−34	−29	−24	−19	−14	−9	−4	1	6	11	16

b. Vertical Intercept: (0, −9)   Horizontal Intercept: $\left(\dfrac{9}{5}, 0\right)$

c. The graph as produced on a TI-83 with is shown.

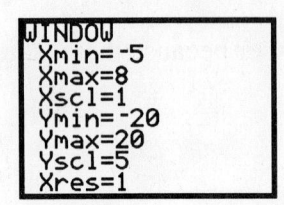

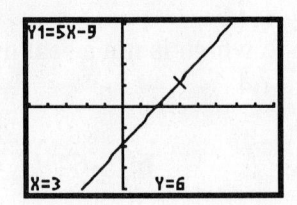

```
WINDOW
Xmin=-5
Xmax=8
Xscl=1
Ymin=-20
Ymax=20
Yscl=5
Xres=1
```

Y1=5X−9, X=3, Y=6

d. The function is increasing. As the x-coordinates increase the y-coordinates also increase.

10. a. **Table 2: Input-Output Table**

x	−5	−4	−3	−2	−1	0	1	2	3	4	5
$y(x)$	20	9	0	−7	−12	−15	−16	−15	−12	−7	0

b. Vertical Intercept: (0, −15)   Horizontal Intercepts: (−3, 0) and (5, 0)

c. The graph is shown as it appears on a TI-83. The vertex has been highlighted using the Trace item in the calculate menu and setting the value of x to be 1.

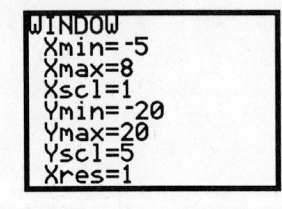

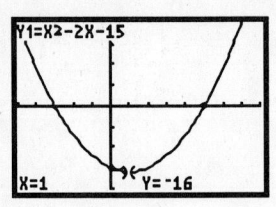

```
WINDOW
Xmin=-5
Xmax=8
Xscl=1
Ymin=-20
Ymax=20
Yscl=5
Xres=1
```

Y1=X²−2X−15, X=1, Y=−16

d. The function is neither increasing nor decreasing. As the x-coordinates increase the y-coordinates decrease until x is 1 and then begin to increase.

11. a. Slope = 4   b. output at the vertical intercept: −11

c. input at the horizontal intercept: $\dfrac{11}{4}$

d. The function is increasing. The graph is shown with the horizontal intercept point shown.

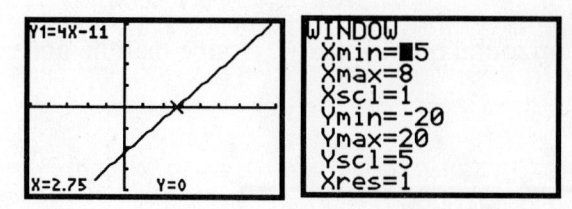

12. a. Slope = $\frac{2}{7}$ b. output at the vertical intercept: 5

    c. input at the horizontal intercept: $-\frac{35}{2}$

    d. The function is increasing. The graph is shown with the horizontal intercept highlighted.

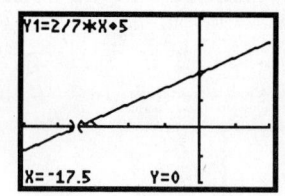

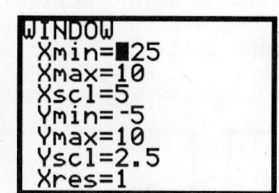

13. a. Slope = $-\frac{3}{5}$ b. output at the vertical intercept: $\frac{5}{4}$

    c. input at the horizontal intercept: $\frac{25}{12}$

    d. The function is decreasing. The graph is shown with the horizontal intercept highlighted.

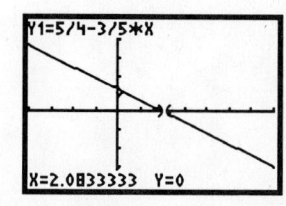

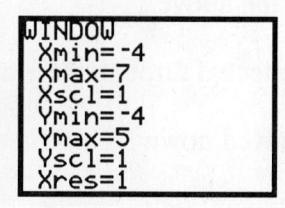

14. a. Slope = 0 b. output at the vertical intercept: 0

    c. All real numbers. d. The graph is the horizontal axis.

15. a.

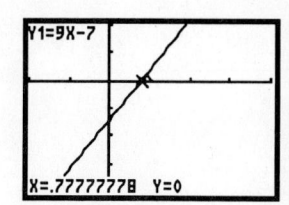

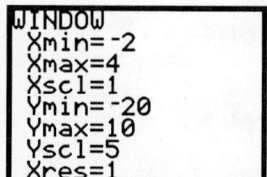

127

b. $y(x) = 9x - 7$

c. The input at the horizontal intercept is $\frac{7}{9}$. Notice that the horizontal intercept is shown on the graph.

16. a.

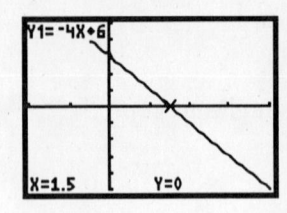

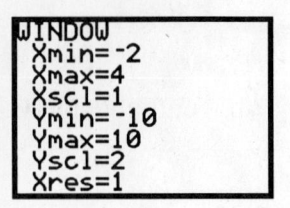

b. $y(x) = -4x + 6$

c. The input at the horizontal intercept is $\frac{3}{2}$. Notice that the horizontal intercept is shown on the above graph.

17. a.

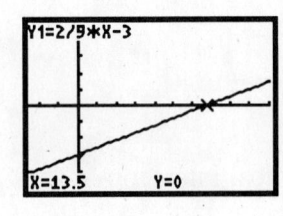

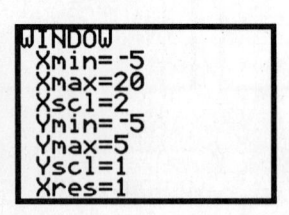

b. $y(x) = \frac{2}{9}x - 3$

c. The input at the horizontal intercept is $\frac{27}{2}$. Notice that the horizontal intercept is shown on the graph above.

18. a. Narrower and reflected through the x-axis.

b. Narrower and moved down 3 units.

c. Wider and moved up 7 units.

d. Narrower, reflected through the x-axis and moved down 4 units.

19. a. Zeros: 5, −5   x-value of vertex: 0

b. Zeros: 2, −7   x-value of vertex: $-\frac{5}{2}$

c. Zeros: −1, −9   x-value of vertex: −5

d. Zeros: 6, 3 x-value of vertex: $\dfrac{9}{2}$

20. a. $y(x) = (x-4)(x+7) = x^2 + 3x - 28$

A display of a table of values of both expressions is shown.

b. $y(s) = (x+8) \cdot (x+6) = x^2 + 14x + 48$

The table values shown computed from each expression are the same.

c. $y(x) = (x-5) \cdot (x+5) = x^2 - 25$

d. $y(x) = (x+6)^2 = x^2 + 12x + 36$

21. a. $-6x^3 - 7x^2 + 7x + 3$ b. $35x^3y^5 - 21xy^6$

c. $16x^3 - 36x^2y - 29xy^2 + 35y^3$ d. $4a^2 - 9b^2$

e. $4x^2 - 28x + 49$ f. $25w^2 + 80wz + 64z^2$

22. a. The terms are $4x$ and -11.   b. The numerical coefficients are 4 and -11.

23. a. The terms are $5x^2$, $-7x$ and -8. b. The numerical coefficients are 5, -7 and -8.

24. a. $-2x^5 + 8x^4 - x^3 + x^2 + 11x - 3$

25. a. $\dfrac{x}{4}$ would be a linear function which would have outputs which are the length of the square.

b. The area of the square would be given by the quadratic function $\left(\dfrac{x}{4}\right)^2 = \dfrac{x^2}{16}$.

CHAPTER 7

Answering Questions with Linear and Quadratic Functions

Section 7.1 Linear Equations and Inequalities in One Variable

p. 551

2.   a.   2   b. -5

      c. $\dfrac{8}{7}$ d. $-\dfrac{4}{3}$

      e. $-\dfrac{2.7}{4.3} = -\dfrac{27}{43}$ f. $-\dfrac{-4}{\tfrac{3}{5}} = 4\left(\dfrac{5}{3}\right) = \dfrac{20}{3}$

3.   a. $x - 2 = 0$ b. $x + 5 = 0$
            $x - 2 + 2 = 0 + 2$ $x + 5 - 5 = 0 - 5$
                 $x = 2$ $x = -5$

      c. $7t - 8 = 0$ d. $3r + 4 = 0$
            $7t - 8 + 8 = 0 + 8$ $3r + 4 - 4 = 0 - 4$
                 $7t = 8$ $3r = -4$
                 $\dfrac{7t}{7} = \dfrac{8}{7}$ $\dfrac{3r}{3} = \dfrac{-4}{3}$
                 $t = \dfrac{8}{7}$ $r = -\dfrac{4}{3}$

e. $$4.3z + 2.7 = 0$$
$$4.3z + 2.7 - 2.7 = 0 - 2.7$$
$$4.3z = -2.7$$
$$\frac{4.3z}{4.3} = \frac{-2.7}{4.3}$$
$$z = -\frac{27}{43}$$

f. $$\frac{3}{5}x - 4 = 0$$
$$\frac{3}{5}x - 4 + 4 = 0 + 4$$
$$\frac{3}{5}x = 4$$
$$\frac{\frac{3}{5}x}{\frac{3}{5}} = \frac{4}{\frac{3}{5}}$$
$$x = 4\left(\frac{5}{3}\right)$$
$$x = \frac{20}{3}$$

4. a. $$5x - 3 = 2$$
$$5x - 3 + 3 = 2 + 3$$
$$5x = 5$$
$$\frac{5x}{5} = \frac{5}{5}$$
$$x = 1$$

b. $$x + 7 = 4$$
$$x + 7 - 7 = 4 - 7$$
$$x = -3$$

c. $$3x - 13 = -27$$
$$3x - 13 + 13 = -27 + 13$$
$$3x = -14$$
$$\frac{3x}{3} = \frac{-14}{3}$$
$$x = -\frac{14}{3}$$

d. $$5.2x + 11 = 4$$
$$5.2x + 11 - 11 = 4 - 11$$
$$5.2x = -7$$
$$x = \frac{-7}{5.2} = -\frac{70}{52} = -\frac{35}{26}$$

e. $-\frac{2}{3}x - 5 = -\frac{1}{4}$

$\frac{2}{3}x - 5 + 5 = -\frac{1}{4} + 5$

$-\frac{2}{3}x = -\frac{1}{4} + \frac{20}{4}$

$-\frac{2}{3}x = \frac{19}{4}$

$\dfrac{-\frac{2}{3}x}{-\frac{2}{3}} = \dfrac{\frac{19}{4}}{-\frac{2}{3}}$

$x = \left(\frac{19}{4}\right)\left(-\frac{3}{2}\right)$

$x = -\frac{57}{8}$

5. a. $5x - 3 < 2$
    $5x - 3 + 3 < 2 + 3$
    $5x < 5$
    $\frac{5x}{5} < \frac{5}{5}$
    $x < 1$

    b. $x + 7 > 4$
    $x + 7 - 7 > 4 - 7$
    $x > -3$

    c. $3x - 13 \leq -27$
    $3x - 13 + 13 \leq -27 + 13$
    $3x \leq -14$
    $\frac{3x}{3} \leq \frac{-14}{3}$
    $x \leq -\frac{14}{3}$

    d. $5.2x + 11 \geq 4$
    $5.2x + 11 - 1 \geq 4 - 11$
    $5.2x \geq -7$
    $\frac{5.2x}{5.2} \geq \frac{-7}{5.2}$
    $x \geq -\frac{35}{26}$

e. $-\frac{2}{3}x - 5 < -\frac{1}{4}$

$-\frac{2}{3}x - 5 + 5 < -\frac{1}{4} + 5$

$-\frac{2}{3}x < -\frac{1}{4} + \frac{20}{4}$

$-\frac{2}{3}x < \frac{19}{4}$

$\dfrac{-\frac{2}{3}x}{-\frac{2}{3}} > \dfrac{\frac{19}{4}}{-\frac{2}{3}}$

$x > \left(\frac{19}{4}\right)\left(-\frac{3}{2}\right)$

$x > -\frac{57}{8}$

6. a. $4x + y = -13$ Slope $= -4$ Vertical Intercept: $(0, -13)$
$4x + y - 4x = -13 - 4x$
$y = -4x - 13$

b. $2x - 9y = 17$ Slope $= \frac{2}{9}$ Vertical Intercept: $\left(0, -\frac{17}{9}\right)$
$2x - 9y - 2x = 17 - 2x$
$-9y = -2x + 17$
$\frac{-9y}{-9} = \frac{-2x + 17}{-9}$
$y = \frac{2}{9}x - \frac{17}{9}$

c. $x + 5y = 0$ Slope $= -\frac{1}{5}$ Vertical Intercept: $(0, 0)$
$x + 5y - x = 0 - x$
$5y = -x$
$y = -\frac{1}{5}x$

d. $4x - 3y = -1$ Slope = $\dfrac{4}{3}$ Vertical Intercept: $\left(0, \dfrac{4}{3}\right)$
$4x - 3y - 4x = -1 - 4x$
$-3y = -4x - 1$
$\dfrac{-3y}{-3} = \dfrac{-4x - 1}{-3}$
$y = \dfrac{4}{3}x + \dfrac{1}{3}$

7. The equation must be of the form $y = -5x + b$. So, if we substitute -3 for x and 1 for y we can solve the resulting equation for b. Solving $1 = -5(-3) + b$ gives us $b = -14$. The requested function is defined by the equation $y(x) = -5x - 14$.

8. a. The slope of the line determined by the given points is $\dfrac{7 - 3}{4 - (-6)} = \dfrac{4}{10} = \dfrac{2}{5}$. The equation must be of the form $y = \dfrac{2}{5}x + b$. So, we solve the equation we get when we substitute 4 for x and 7 for y. $7 = \dfrac{2}{5} \cdot 4 + b$ gives us $b = \dfrac{27}{5}$. The linear function is given by the equation $y(x) = \dfrac{2}{5}x + \dfrac{27}{5}$.

    b. Solving $-3 = \dfrac{2}{5} \cdot 1 + b$ gives us $b = -\dfrac{17}{5}$. The requested equation is $y(x) = \dfrac{2}{5}x - \dfrac{17}{5}$.

    c. The equation would be of the form $y = -\dfrac{5}{2}x + b$. So, we must solve $-3 = -\dfrac{5}{2} \cdot 1 + b$ to find b. We get $b = -\dfrac{1}{2}$. The requested equation is $y(s) = -\dfrac{5}{2}x - \dfrac{1}{2}$.

9. a. First solve $2x + 3y = 4$ for y. The slope is the coefficient of x in that equation. $3y = -2x + 4$ $y = -\dfrac{2}{3}x + \dfrac{4}{3}$ The slope is $-\dfrac{2}{3}$. The requested equation must contain the point $(-3, 5)$ so, we solve $5 = -\dfrac{2}{3}(-3) + b$ to get $b = 3$. The requested equation is $y = -\dfrac{2}{3}x + 3$.

134

b. In this equation the slope should be $\frac{3}{2}$, so we solve $5 = \frac{3}{2}(-3) + b$ to get $b = \frac{19}{2}$.

The requested equation is $y(x) = \frac{3}{2}x + \frac{19}{2}$.

10. a. $T(m) = 0.2m + 15$

   b.
   $$T(m) = 40$$
   $$0.2m + 15 = 40$$
   $$0.2m + 15 - 15 = 40 - 15$$
   $$0.2m = 25$$
   $$\frac{0.2m}{0.2} = \frac{25}{0.2}$$
   $$m = 125$$

   125 miles were driven for a total charge of $40.

   c.
   $$T(m) = 100$$
   $$0.2m + 15 = 100$$
   $$0.2m + 15 - 15 = 100 - 15$$
   $$0.2m = 85$$
   $$\frac{0.2m}{0.2} = \frac{85}{0.2}$$
   $$m = 425$$

   425 miles were driven for a total charge of $100.

11. a.
   $$0.25c - 30 = 75$$
   $$0.25c - 30 + 30 = 75 + 30$$
   $$0.25c = 105$$
   $$\frac{0.25c}{0.25} = \frac{105}{0.25}$$
   $$c = 420$$

   420 cookies must be sold to realize a profit of $75.

   b.
   $$0.25c - 30 = 100$$
   $$0.25c - 30 + 30 = 100 + 30$$
   $$0.25c = 130$$
   $$\frac{0.25c}{0.25} = \frac{130}{0.25}$$
   $$c = 520$$

   520 cookies must be sold to realize a profit of $100.

12. $y(x) = 5x + 2$

    a. $5x + 2 = 14$
        $5x = 12$
        $x = \dfrac{12}{5}$

    b. $5x + 2 = -4$
        $5x = -6$
        $x = \dfrac{-6}{5}$

    c. $5x + 2 \leq 2$
        $5x \leq 0$
        $x \leq 0$

    d. $5x + 2 > 4$
        $5x > 2$
        $x > \dfrac{2}{5}$

13. $y(x) = 3 - 2x$

    a. $3 - 2x = 0$
        $-2x = -3$
        $x = \dfrac{3}{2}$

    b. $3 - 2x = 7$
        $-2x = 4$
        $x = -2$

    c. $3 - 2x < 3$
        $-2x < 0$
        $x > 0$

    When multiplying or dividing an inequality by a negative number, the sense of the inequality reverses (Compare step 2 to step 3).

    d. $3 - 2x \geq 4$
        $-2x \geq 1$
        $x \leq -\dfrac{1}{2}$

    When multiplying or dividing an inequality by a negative number, the sense of the inequality reverses (Compare step 2 to step 3).

14. $y(x) = 7x + 2$

    a. $y(3) = 7(3) + 2 = 23$

    b. $7x + 2 = 3$
        $7x = 1$
        $x = \dfrac{1}{7}$

15. a. $y(1) \approx -3$ b. $y(-3) \approx -14$

 c. $y(x) = 0 \Rightarrow x \approx 2$ d. $y(x) = 9 \Rightarrow x \approx 5$

 e. It appears that $y(x) = 5$ when $x \approx 3.5$. Since the outputs are increasing as the input increases, $y(x) > 5$ when $x > 3.5$.

 f. It appears that $y(x) = -1$ when $x \approx 1.5$. Since the outputs are increasing as the input increases, $y(x) \leq -1$ when $x \leq 1.5$.

Section 7.2 Systems of Equations

p. 569

2. a. $\begin{array}{l} x + y = 8 \\ x - y = 4 \end{array}$ Solve $x + y = 8$ for x by subtracting y from both sides. The equation becomes $x = 8 - y$. Substitute this for x in $x - y = 4$.

We get	$(8 - y) - y = 4$
Simplifying the left side:	$8 - 2y = 4$
Solve for y	
Subtract 8 from both sides	$-2y = -4$
Divide both sides by -2	$y = 2$
Substitute into $x = 8 - y$	$x = 8 - 2 = 6$

The solution to the system is $(6, 2)$.

 b. $\begin{array}{l} x + 2y = 5 \\ x + y = 3 \end{array}$. Solve $x + y = 3$ for x by subtracting y from both sides. The equation becomes $x = 3 - y$. Substitute this for x in $x + 2y = 5$.

We get	$(3 - y) + 2y = 5$
Simplifying the left side	$3 + y = 5$
Solve for y	
Subtract 3 from both sides	$y = 2$
Substitute into $x = 3 - y$	$x = 3 - 2 = 1$ The solution to the system is $(1, 2)$.

c. $\begin{array}{l}5x - y = 13\\2x + 3y = 12\end{array}$. Solve $5x - y = 13$ for y by adding y to both sides to get $5x = 13 + y$. Then subtract 13 from both sides to get $5x - 13 = y$. Substitute for y in $2x + 3y = 12$.

We get	$2x + 3(5x - 13) = 12$
Simplify the left side	$2x + 15x - 39 = 12$
	$17x - 39 = 12$

Solve for x

Add 39 to both sides	$17x = 51$
Divide both sides by 17	$x = 3$
Substitute into $5x - 13 = y$	$y = 5(3) - 13 = 15 - 13 = 2$

The solution to the system is $(3, 2)$.

d. $\begin{array}{l}x + 3y = 5\\2x - 3y = -8\end{array}$. Solve $x + 3y = 5$ for x by subtracting $3y$ from both sides to obtain $x = 5 - 3y$. Substitute for x in $2x - 3y = -8$

We get	$2(5 - 3y) - 3y = -8$
Simplify the left side	$10 - 6y - 3y = -8$
	$10 - 9y = -8$
Subtract 10 from both sides	$-9y = -18$
Divide both sides by -9	$y = 2$
Substitute into $x = 5 - 3y$	$x = 5 - 3(2) = 5 - 6 = -1$

The solution to the system is $(-1, 2)$.

e. $\begin{array}{l}3x - 2y = 0\\5x + 10y = 4\end{array}$. Solve $3x - 2y = 0$ for y. Add $2y$ to both sides obtaining $3x = 2y$.

Divide both sides by 2 obtaining $y = \dfrac{3}{2}x$. Substitute for y into $5x + 10y = 4$.

138

We get $\qquad 5x + 10\left(\dfrac{3}{2}x\right) = 4$

Simplify the left side $\qquad 5x + 15x = 4$

$\qquad\qquad\qquad\qquad\qquad\qquad 20x = 4$

Divide both sides by 20 $\qquad x = \dfrac{1}{5}$

Substitute into $y = \dfrac{3}{2}x$ $\qquad y = \left(\dfrac{3}{2}\right)\left(\dfrac{1}{5}\right) = \dfrac{3}{10}$

The solution to the system is $\left(\dfrac{1}{5}, \dfrac{3}{10}\right)$.

f. The second equation is obtained from the first by multiplying both sides of the first by two. These represent the same relationship and the solution to the system is all ordered pairs that satisfy the relationship.

g. There are no ordered pairs in the solution to the system. The graphs of the two equations do not intersect since they are parallel lines.

h. $\begin{array}{l} 5x + y = -2 \\ 2x + 7y = 3 \end{array}$ Solve the first equation for y. We get $y = -5x - 2$. Substitute this expression for y in the second equation to get $2x + 7(-5x - 2) = 3$. Solving this equation gives us $2x - 35x - 14 = 3$ which is equivalent to $-33x - 14 = 3$. Adding 14 to both sides and then dividing by -33 gives $x = -\dfrac{17}{33}$. Substitute this value for x in the first equation where we solved for y to get

$y = -5\left(-\dfrac{17}{33}\right) - 2 = \dfrac{85}{33} - 2 = \dfrac{85}{33} - \dfrac{66}{33} = -\dfrac{19}{33}$. The solution is $\left(-\dfrac{17}{33}, -\dfrac{19}{33}\right)$.

3. The equations in 1f are dependent since they represent the same relationship.

4. The equations in 1g are inconsistent since they have no common solutions. The system in 1f has an infinite number of solutions while the system in 1g has no solutions.

5. Answer is based on each person's individual experience.

6. a. The graphs do not intersect since the slopes are the same and the horizontal intercepts are different.

   b. There are no solutions to this system of equations because the two graphs do not intersect. We could say that the solution set is the empty set.

c. The graphs of your two equations should have the same slope and different horizontal intercepts.

7. a.
$$2x - 3 = x + 4$$
$$2x - x - 3 + 3 = x - x + 4 + 3$$
$$x = 7$$

b.
$$7x + 2 = 4x - 5$$
$$7x - 4x + 2 - 2 = 4x - 4x - 5 - 2$$
$$3x = -7$$
$$x = -\frac{7}{3}$$

c.
Original inequality	$x - 8 > 2 - 3x$
Add 3x to both sides	$x - 8 + 3x > 2 - 3x + 3x$
Simplify	$4x - 8 > 2$
Add 8 to both sides	$4x - 8 + 8 > 2 + 8$
Simplify	$4x > 10$
Divide both sides by 4	$\frac{4x}{4} > \frac{10}{4}$
Simplify	$x > \frac{5}{2}$

d.
Original inequality	$5x + 4 \le 3x$
Subtract 5x from both sides	$5x + 4 - 5x \le 3x - 5x$
Simplify	$4 \le -2x$
Divide both sides by –2 and reverse the direction of the inequality.	$\frac{4}{-2} \ge \frac{-2x}{-2}$
Simplify	$-2 \ge x$

e. $6t + 11 = 6 - 5t$
    $11t = -5$
    $t = -\dfrac{5}{11}$

f. $2 - z = 11z + 7$
    $-5 = 12z$
    $z = -\dfrac{5}{12}$

g. $4k - 9 < 7k + 2$
    $-11 < 3k$
    $-\dfrac{11}{3} < k$

h. $9 - 5p < 2p - 3$
    $12 < 7p$
    $\dfrac{12}{7} < p$

i. $3z + 1 < 2 - 7z$
    $10z < 1$
    $z < \dfrac{1}{10}$

j. $a - 1 < 4a + 1$
    $-2 < 3a$
    $-\dfrac{2}{3} < a$

8. In the equation $8B = 40$, B represents the number of books, not the price of each book.

9. The system for the basket ball problem is $\begin{array}{l} 2x + 3y = 58 \\ x + y = 27 \end{array}$ where x is the number of two–point shots made and y is the number of three point shots made. The numerical and graphical solutions appear below as displayed on a TI-82.

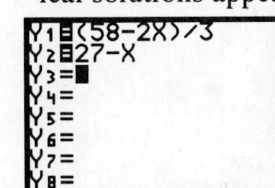

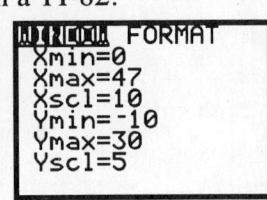

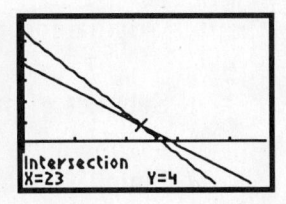

Solving algebraically, solve the second equation for y.

$$y = 27 - x$$

Substitute $(27 - x)$ for y in the first equation $2x + 3(27 - x) = 58$

Simplify left side $2x + 81 - 3x = 58$

$-x + 81 = 58$

Subtract 81 from both sides $-x + 81 - 81 = 58 - 81$

Simplify $-x = -23$

Divide both sides by -1. $\dfrac{-x}{-1} = \dfrac{-23}{-1}$

141

Simplify $\qquad$ $x = 23$

If $x = 23$ then $y = 27 - 23 = 4$

Thus, the Bulls made 23 two–point shots and 4 three–point shots.

10. a. $2x - 3 = x + 4$

   b. Original equation $\qquad$ $2x - 3 = x + 4$

   Subtract x from both sides $\qquad$ $2x - 3 - x = x + 4 - x$

   Simplify both sides $\qquad$ $x - 3 = 4$

   Add 3 to both sides $\qquad$ $x - 3 + 3 = 4 + 3$

   Simplify both sides $\qquad$ $x = 7$

   The outputs of the two functions are equal when the input is 7.

   c. $2x - 3 > x + 4$

   d. Original inequality $\qquad$ $2x - 3 > x + 4$

   Subtract x from both sides $\qquad$ $2x - 3 - x > x + 4 - x$

   Simplify both sides $\qquad$ $x - 3 > 4$

   Add 3 to both sides $\qquad$ $x - 3 + 3 > 4 + 3$

   Simplify both sides $\qquad$ $x > 7$

   The output of $y(x) = 2x - 3$ exceeds the output of $y(x) = x + 4$ when the input exceeds 7.

11. a. Equation $\qquad$ $3x - 7 = 4 - 5x$

   b. Add $5x$ to both sides $\qquad$ $3x - 7 + 5x = 4 - 5x + 5x$

   Simplify both sides $\qquad$ $8x - 7 = 4$

   Add 7 to both sides $\qquad$ $8x - 7 + 7 = 4 + 7$

   Simplify both sides $\qquad$ $8x = 11$

Divide both sides by 8 $\qquad \dfrac{8x}{8} = \dfrac{11}{8}$

Simplify both sides $\qquad x = \dfrac{11}{8}$

So the outputs are equal when the input is $\dfrac{11}{8}$.

c. Inequality $\qquad 3x - 7 \leq 4 - 5x$

d. We use the same steps as above to solve the inequality. Since we never multiply or divide both sides by a negative number, the direction of the inequality remains the same throughout the solving process. The output of $y = 3x - 7$ will be no larger than the output of $y = 4 - 5x$ when the input is less than or equal to $\dfrac{11}{8}$.

Section 7.3 Finding Zeros of Quadratic Functions by Factoring

p. 589

2. a. The zeros are –2 and 6.

   b. The factors are $(x - (-2)) = (x + 2)$ and $(x - 6)$.

   c. The function is $y(x) = (x + 2)(x - 6) = x^2 - 4x - 12$.

3. a. We must solve $x^2 + 11x + 12 = -16$. Add –16 to both sides to place zero on the right hand side and get the equation $x^2 + 11x + 28 = 0$. Factor the left side: $(x + 7)(x + 4) = 0$. Now we can see that the solutions are –7 and –4. Using a table of ouputs for the function $y(x) = x^2 + 11x + 12$ we see that the inputs of –7 and –4 do produce an output of –16. The graph shows this too.

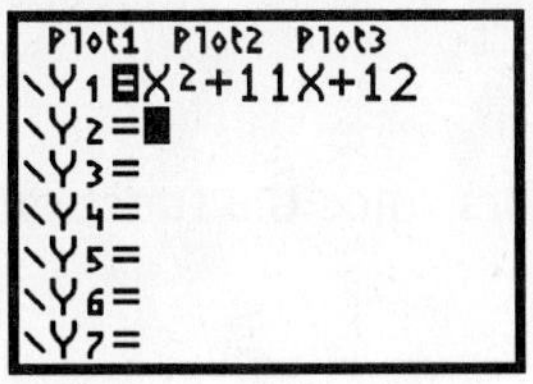

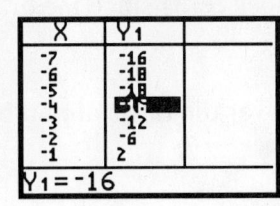

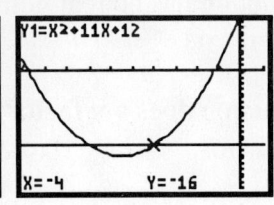

143

b. We must solve the equation $x^2 - 2x + 1 = 16$. First subtract 16 form both sides $x^2 - 2x - 15 = 0$. Factor the left side: $(x-5)(x+3) = 0$. The factors and hence the product will be zero if $x = 5$ or $x = -3$.

If we enter the original function into the Y= menu and produce a table we see that the inputs of 5 and -3 produce an output of 16. The graph shows the solutions too.

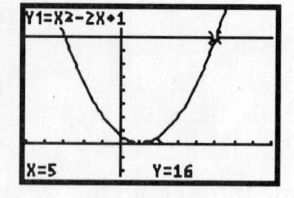

c. We must solve $x^2 - 10x + 10 = -6$. First add 6 to both sides: $x^2 - 10x + 16 = 0$. Now factor the left side: $(x-8)(x-2) = 0$. We can now see that $x = 8$ and $x = 2$ make the two factors and hence the product equal to zero. The table and graph appear below.

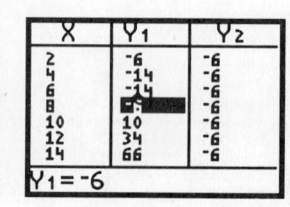

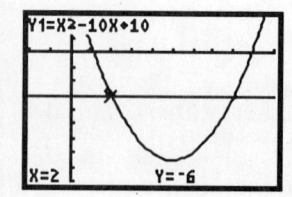

d. Solve the equation $x^2 - 5x + 9 = 9$. Subtract 9 form both sides: $x^2 - 5x = 0$. Factor the left hand side: $x(x-5) = 0$. We see that $x = 0$ and $x = 5$ make the factors and hence the product equal to zero. The table and graph are displayed below as a check.

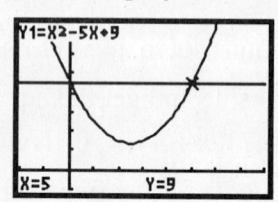

4. a. This function has no real number zeros since the graph does not intersect the horizontal axis.

    b. This function does not factor over the rational numbers since the function has no real number zeros.

5. a. Zeros are 0 and 2. $9x^2 - 18x = 9x(x-2)$

    b. Zeros are 0 and $\frac{2}{3}$. $21y^4 - 14y^3 = 7y^3(3y-2)$

144

c. Zeros are 0, –6, and 1. $5x^3 + 25x^2 - 30x = 5x(x^2 + 5x - 6) = 5x(x+6)(x-1)$

d. Zeros are –1 and –8. $\qquad x^2 + 9x + 8 = (x+1)(x+8)$

e. Zeros are 7 and –1. $\qquad p^2 - 6p - 7 = (p-7)(p+1)$

f. No rational number zeros. $\qquad x^2 + 12x + 13$ is prime.

g. Zeros are 6 and –5. $\qquad x^2 - x - 30 = (x-6)(x+5)$

h. No rational number zeros. $\qquad y^2 - 2y - 45$ is prime.

i. Zeros are –5 and –3. $\qquad x^2 + 8x + 15 = (x+5)(x+3)$

j. Zeros are 4 and 2. $\qquad x^2 - 6x + 8 = (x-4)(x-2)$

k. Zeros are –11 and 2. $\qquad x^2 + 9x - 22 = (x+11)(x-2)$

l. Zeros are –5 and 3. $\qquad p^2 + 2p - 15 = (p+5)(p-3)$

m. Zeros are 8 and –7. $\qquad x^2 - x - 56 = (x-8)(x+7)$

n. Zeros are 4 and 11. $\qquad x^2 - 15x + 44 = (x-4)(x-11)$

o. Zero is –7. $\qquad x^2 + 14x + 49 = (x+7)(x+7)$

p. Zero is 6. $\qquad y^2 - 12y + 36 = (y-6)(y-6)$

6. a. $x^2 + 3x = 70$.

   Make right side zero by subtracting 70 from both sides: $x^2 + 3x - 70 = 0$.

   Factor: $(x+10)(x-7) = 0$.

   The zeros are –10 and 7. So the width is 7 yards and the length is 10 yards.

   b. $x^2 + 3x = 108$.

   Make right side zero by subtracting 108 from both sides: $x^2 + 3x - 108 = 0$.

   Factor: $(x+12)(x-9) = 0$.

The zeros are −12 and 9. So the width is 9 yards and the length is 12 yards.

7. a. $x + 7$
   b. $A(x) = x(x+7) = x^2 + 7x$

   c. $A(9) = 9^2 + 7(9) = 81 + 63 = 144$. So the area is 144 square feet when the width is 9 feet.

   d. If the length is 12 feet then the width is 5 feet.

   So $A(5) = 5^2 + 7(5) = 25 + 35 = 60$. The area is 60 square feet when the length is 12 feet.

   e. $A(x) = 78$. So $x^2 + 7x = 78$. Make the right side zero by subtracting 78 from both sides to get $x^2 + 7x - 78 = 0$. Factor: $(x+13)(x-6) = 0$. The zeros are −13 and 6. So the width is 6 feet and the length is 13 feet when the area is 78 square feet.

8. a. $\dfrac{x^2 - 8x + 15}{x^2 - 2x - 15} = \dfrac{(x-5)(x-3)}{(x-5)(x+3)} = \dfrac{x-3}{x+3}$

   b. $\dfrac{x^2 + 9x + 14}{x^2 + 2x - 35} = \dfrac{(x+2)(x+7)}{(x+7)(x-5)} = \dfrac{x+2}{x-5}$

   c. $\dfrac{3x - 6}{x^2 - x - 2} = \dfrac{3(x-2)}{(x-2)(x+1)} = \dfrac{3}{x+1}$

   d. Since the denominator cannot be factored the fraction is in reduced form, as is.

9. a. $\dfrac{5}{2x-10} - \dfrac{7}{3x-15} = \dfrac{5}{2(x-5)} - \dfrac{7}{3(x-5)} = \dfrac{15}{6(x-5)} - \dfrac{14}{6(x-5)} = \dfrac{1}{6(x-5)}$

   b. $\dfrac{2x}{(x-2)(x+1)} - \dfrac{5}{(x+3)(x+1)} = \dfrac{2x(x+3)}{(x-2)(x+1)(x+3)} - \dfrac{5(x-2)}{(x+3)(x+1)(x-2)} =$

   $\dfrac{2x(x+3) - 5(x-2)}{(x-2)(x+1)(x+3)} = \dfrac{2x^2 + 6x - 5x + 10}{(x-2)(x+1)(x+3)} = \dfrac{2x^2 + x + 10}{(x-2)(x+1)(x+3)}$

10. a. $5x + 7 = 11x + 5$
    $2 = 6x$
    $x = \dfrac{2}{6} = \dfrac{1}{3}$

    b. $5x + 7 > 11x + 5$
    $2 > 6x$
    $\dfrac{1}{3} > x$

c.
$5x + 7 < 11x + 5$
$2 < 6x$
$\frac{1}{3} < x$

d.
$7 - 3t = 8t + 6$
$1 = 11t$
$\frac{1}{11} = t$

e.
$3x - 11 < 5x - 1$
$-2x < 10$
$x > -5$

f.
$9x > 3 - 4x$
$13x > 3$
$x > \frac{3}{13}$

g. $7 - t = 4 - t$ has no solution.

h.
$5 + 3x = 4 - 7x$
$10x = -1$
$x = -\frac{1}{10}$

i.
$\frac{2}{3}z - \frac{5}{6} = 4z + \frac{3}{4}$
$12\left(\frac{2}{3}z - \frac{5}{6}\right) = 12\left(4z + \frac{3}{4}\right)$
$8z - 10 = 48z + 9$
$-19 = 40z$
$-\frac{19}{40} = z$

j.
$1 + \frac{1}{2}x = 5x - \frac{2}{3}$
$6\left(1 + \frac{1}{2}x\right) = 6\left(5x - \frac{2}{3}\right)$
$6 + 3x = 30x - 4$
$10 = 27x$
$\frac{10}{27} = x$

11. a. $\begin{array}{l}2x + y = 7 \\ 5x - 2y = -1\end{array}$ First solve the first equation for y to get $y = -2x + 7$. Substitute in the second equation and solve for x.

$5x - 2(-2x + 7) = -1$
$5x + 4x - 14 = -1$
$9x - 14 = -1$
$9x = 13$
$x = \frac{13}{9}$

Substitute this value for x in the first equation and solve for y.

$y = -2\left(\frac{13}{9}\right) + 7 = -\frac{26}{9} + 7 = -\frac{26}{9} + \frac{63}{9} = \frac{37}{9}$ The solution to the system is $\left(\frac{13}{9}, \frac{37}{9}\right)$.

b. $\begin{array}{l}y = 5x - 2 \\ 5x - y = 17\end{array}$ This system has no solution. Each of the equations has a graph which is a straight line with slope 5. But the horizontal intercept of the two graphs are different. Therefore, the graphs are parallel lines, which means that they do not intersect.

c. $\begin{array}{l}3x+4y = 9\\ x-5y = 0\end{array}$ First solve the second equation for x to get $x = 5y$. Substitute into the first equation and solve for y.

$$3(5y)+4y = 9$$
$$19y = 9$$
$$y = \frac{9}{19}$$

and $x = 5\left(\frac{9}{19}\right) = \frac{45}{19}$. The solution to the system is $\left(\frac{45}{19}, \frac{9}{19}\right)$.

d. $\begin{array}{l}x = 2-4y\\ 3x+y = 7\end{array}$ Replace x in the second equation with $2-4y$ and solve for y.

$$3(2-4y)+y = 7$$
$$6-12y+y = 7$$
$$-11y = 1$$
$$y = -\frac{1}{11}$$

And $x = 2-4\left(-\frac{1}{11}\right) = 2+\frac{4}{11} = \frac{26}{11} = 2\frac{4}{11}$. The solution to the system is $\left(\frac{26}{11}, -\frac{1}{11}\right)$.

12. Your system should be of the form $\begin{array}{l}ax+by = c\\ ax+by = d\end{array}$ where c and d are different numbers.

13. We must solve $3x-2 = 7x+9$ first. The solution is $x = -\frac{11}{4}$. To find the y-coordinate find the output for either function using this value as the input.

$y = 3\left(-\frac{11}{4}\right)-2 = -\frac{33}{4}-\frac{8}{4} = -\frac{41}{4}$ The intersection point is $\left(-\frac{11}{4}, -\frac{41}{4}\right)$.

14. a. $x+2y = 7$ is equivalent to $y = -\frac{1}{2}x+\frac{7}{2}$. So, the slope of the given line is $-\frac{1}{2}$. The equation of the requested line must be of the form $y = -\frac{1}{2}x+b$. To find b we substitute the coordinates $(-1, 2)$ into the equation and solve for b. $2 = -\frac{1}{2}(-1)+b$ has the solution $b = \frac{3}{2}$. The requested equation is $y = -\frac{1}{2}x+\frac{3}{2}$.

b. $4x - 3y = -1$ is equivalent to $y = \frac{4}{3}x + \frac{1}{3}$ so, the slope of the given line is $\frac{4}{3}$. The slope of a line perpendicular to this line must be $-\frac{3}{4}$. Solve $0 = -\frac{3}{4}(-4) + b$ to get $b = -3$. The equation of the requested line is $y = -\frac{3}{4}x - 3$.

Section 7.4 Additional Factoring Experiences

p. 607

2.   a. prime   b. $(2x+1)(3x-1)$

    c. $10(4y^2 + 3)$ d. $(x+8)(x+5)$

    e. prime   f. $(y-2)(2y-1)$

    g. $(y-3)(8y+7)$ h. $(2x+3)(3x-2)$

    i. $(x+4)(3x-1)$ j. $(5y-11)(5y+11)$

    k. prime   l. prime

    m. $6(3x+2)^2$ n. $(x+2)(x+7)$

    o. $(3a+5)(2a-1)$ p. $(x+3)(x-9)$

    q. $2(y-4)(2y-1)$ r. $(x+4)(4x-1)$

    s. $t(t-11)$ t. $(2x+5)(3x-7)$

    u. $(2x-5)(4x+1)$ v. $(x-5)(4x+9)$

    w. $(x+8)^2$ x. $(3x-1)(3x+1)$

3.   a. $x+5$ b. $A(x) = x(x+5) = x^2 + 5x$

    c. $A(11) = 11(11+5) = 11(16) = 176$ The area is 176 square feet when the width is 11 feet.

    d. We must solve $x(x+5) = 14$. Simplifying the left side and subtracting 14 from both sides we get $x^2 + 5x - 14 = 0$. Factoring the left side we get $(x+7)(x-2) = 0$. The two solutions to this equation are -7 and 2. 2 is the only one that makes sense in this problem. So, the requested width is 2 feet and the length is 7 feet.

e. Solve this in a way similar to part d. 4 feet would be the width and 9 feet would be the length. Notice that the length is 5 feet more than the width and the area would be 36 square feet.

4.   a. $2x - 1$ b. $A(x) = x(2x - 1)$

      c. $A(5) = 5(2 \cdot 5 - 1) = 5 \cdot 9 = 45$ The area would be 45 square feet.

      d.   Solve $x(2x - 1) = 1$. The equation is equivalent to $2x^2 - x - 1 = 0$. Factoring the left hand side gives us $(2x + 1)(x - 1) = 0$. The only solution which makes sense in the problems $x = 1$. So, the width is 1 foot and the length is also 1 foot.

      e.   Set up the equation $x(2x - 1) = 6$. The only solution to this equation which makes sense is $x = 2$. The width is 2 feet and the length is 3 feet.

5.   **Table 3: Factoring**

Factored Form	Expanded Form
$(x - 2)(x^2 + 2x + 4)$	$x^3 - 8$
$(x + 2)(x^2 - 2x + 4)$	$x^3 + 8$
$(x - 3)(x^2 + 3x + 9)$	$x^3 - 27$
$(x + 3)(x^2 - 3x + 9)$	$x^3 + 27$
$(a - b)(a^2 + ab + b^2)$	$a^3 - b^3$
$(a + b)(a^2 - ab + b^2)$	$a^3 + b^3$

6.   a. $(x - 4)(x^2 + 4x + 8)$ b. $(y + 5)(y^2 - 5y + 25)$

      c. $(3a + b)(9a^2 - 3ab + b^2)$ d. $2(2x - 3y)(4x^2 + 6xy + 9y^2)$

7.   a. $2x - \dfrac{7}{3} = \dfrac{9}{5}$ b. $2 - 7z = 5z + 9$

           $2x = \dfrac{62}{15}$ $-7 = 12z$

           $x = \dfrac{31}{15}$ $-\dfrac{7}{12} = z$

c.
$$x^2 + 3x - 4 = 0$$
$$(x+4)(x-1) = 0$$
$$x = -4 \quad \text{or} \quad x = 1$$

d.
$$4x > 7x - 1$$
$$-3x > -1$$
$$x < \frac{1}{3}$$

e.
$$9y + 5 \le 2y - 7$$
$$7y \le -12$$
$$y \le -\frac{12}{7}$$

f.
$$6x^2 - 11x - 10 = 0$$
$$(2x + 5)(3x + 2) = 0$$
$$x = \frac{5}{2} \quad \text{or} \quad x = -\frac{2}{3}$$

g.
$$\frac{4}{7}x - 1 = \frac{2}{5} + 3x$$
$$35\left(\frac{4}{7}x - 1\right) = 35\left(\frac{2}{5} + 3x\right)$$
$$20x - 35 = 14 + 105x$$
$$-49 = 85x$$
$$-\frac{49}{85} = x$$

h.
$$28y^2 + 43y = -10$$
$$28y^2 + 43y + 10 = 0$$
$$(4y + 5)(7y + 2) = 0$$
$$y = -\frac{5}{4} \quad \text{or} \quad y = -\frac{2}{7}$$

i.
$$3 + 10y \ge 6y - 2$$
$$4y \ge -5$$
$$y \ge -\frac{5}{4}$$

j.
$$4x^2 - 7x + 2 = 2$$
$$4x^2 - 7x = 0$$
$$x(4x - 7) = 0$$
$$x = 0 \text{ or } x = \frac{7}{4}$$

k.
$$4t^2 + 2t - 9 = 2t$$
$$4t^2 - 9 = 0$$
$$(2t - 3)(2t + 3) = 0$$
$$t = \frac{3}{2} \text{ or } t = -\frac{3}{2}$$

l.
$$16x^2 + 24x + 9 = 0$$
$$(4x + 3)^2 = 0$$
$$x = -\frac{3}{4}$$

m.
$$\frac{4}{3}z - \frac{1}{6} \ge \frac{3}{4} - z$$
$$12\left(\frac{4}{3}z - \frac{1}{6}\right) \ge 12\left(\frac{3}{4} - z\right)$$
$$8z - 2 \ge 9 - 12z$$
$$20z \ge 11$$
$$z \ge \frac{11}{20}$$

n.
$$2p^2 + 15 = 13p$$
$$2p^2 - 13p + 15 = 0$$
$$(2p - 3)(p - 5) = 0$$
$$p = \frac{3}{2} \text{ or } p = 5$$

8. a.
$$4x - 7(2x - 3) = 1$$
$$4x - 14x + 21 = 1$$
$$-10x + 21 = 1$$
$$-10x = -20$$
$$x = 2$$

$y = 2(2) - 3 = 1$ The intersection point is $(2, 1)$.

b.
$$y = 5(3y + 2) - 1$$
$$y = 15y + 10 - 1$$
$$-9 = 14y$$
$$-\frac{9}{14} = y$$

$$x = 3\left(-\frac{9}{14}\right) + 2 = -\frac{27}{14} + \frac{28}{14} = \frac{1}{14}$$

The intersection point is $\left(\frac{1}{14}, -\frac{9}{14}\right)$.

c. Solve $2x - y = 7$ for y to get $y = 2x - 7$.

Substitute into the second equation and solve
$$3x + 5(2x - 7) = -2$$
$$3x + 10x - 35 = -2$$
$$13x = 33$$
$$x = \frac{33}{13}$$

Then find the y-coordinate by evaluating $y = 2\left(\frac{33}{13}\right) - 7 = \frac{66}{13} - \frac{91}{13} = -\frac{25}{13}$.

The intersection is at $\left(\frac{33}{13}, -\frac{25}{13}\right)$.

d. $x + 3y = 4$ is equivalent to $x = -3y + 4$. Substituting we solve

$$5(-3y + 4) - 8y = -1$$
$$-15y + 20 - 8y = -1$$
$$-23y = -21$$
$$y = \frac{21}{23}$$

$$x = -3\left(\frac{21}{23}\right) + 4 = -\frac{63}{23} + \frac{92}{23} = \frac{29}{23}$$

The intersection point is $\left(\frac{29}{23}, \frac{21}{23}\right)$.

152

9. a. $\dfrac{2x+8}{x^2+3x-4} = \dfrac{2(x+4)}{(x+4)(x-1)} = \dfrac{2}{x-1}; x \neq -4, x \neq 1$

b. $\dfrac{2x^2+x}{4x^2-1} = \dfrac{x(2x+1)}{(2x+1)(2x-1)} = \dfrac{x}{2x-1}; x \neq -\dfrac{1}{2}, x \neq \dfrac{1}{2}$

c. $\dfrac{x^2-10x+25}{x^2+4x+5}$ is reduced since the denominator cannot be factored.

d. $\dfrac{4x^2-4x-8}{6x^2-8x-8} = \dfrac{4(x-2)(x+1)}{2(3x+2)(x-2)} = \dfrac{2(x+1)}{3x+2}; x \neq 2, x \neq -\dfrac{2}{3}$

10. a. $2x-7y = 9$ is equivalent to $y = \dfrac{2}{7}x - \dfrac{9}{7}$. So the slope is $\dfrac{2}{7}$ and the vertical intercept is $\left(0, -\dfrac{9}{7}\right)$.

b. $x+5y-8 = 0$ is equivalent to $y = -\dfrac{1}{5}x + \dfrac{8}{5}$. The slope is $-\dfrac{1}{5}$ and the vertical intercept is $\left(0, \dfrac{8}{5}\right)$.

11. a. $y = \dfrac{2}{3}x + 1$ is the requested equation.

b. The slope of the line is $\dfrac{7-2}{3-(-1)} = \dfrac{5}{4}$. The equation of the line is of the form $y = \dfrac{5}{4}x + b$. Find b by substituting -1 for x and 2 for y. $2 = \dfrac{5}{4}(-1) + b$ is equivalent to $b = \dfrac{13}{4}$. The requested equation is $y = \dfrac{5}{4}x + \dfrac{13}{4}$.

c. $4x-y = 2$ is equivalent to $y = 4x-2$, so the slope of the given line and the requested line is 4. To find b we solve $4 = 4(-3) + b$ which gives us $b = 16$. The equation of the requested line is $y = 4x + 16$.

d. $x+6y = -5$ is equivalent to $y = -\dfrac{1}{6}x - \dfrac{5}{6}$. The slope of the given line is $-\dfrac{1}{6}$ which makes the slope of any line perpendicular to this line 6. To find b we solve $-2 = 6(-1) + b$ which gives us $b = 4$. The equation of the requested line is $y = 6x + 4$.

12. a. $\dfrac{4}{x} - \dfrac{3}{7x} = \dfrac{28}{7x} - \dfrac{3}{7x} = \dfrac{25}{7x}$

   b. $\dfrac{1}{x-2} + \dfrac{2}{x+3} = \dfrac{x+3}{(x-2)(x+3)} + \dfrac{2(x-2)}{(x-2)(x+3)} = \dfrac{x+3+2(x-2)}{(x-2)(x+3)} =$

   $\dfrac{3x-1}{(x-2)(x+3)}$

   c. $\dfrac{9}{(2x-5)(3x+1)} - \dfrac{2x}{(2x-5)(x-4)} =$

   $\dfrac{9(x-4)}{(2x-5)(3x+1)(x-4)} - \dfrac{2x(3x+1)}{(2x-5)(x-4)(3x+1)} =$

   $\dfrac{9(x-4) - 2x(3x+1)}{(2x-5)(3x+1)(x-4)} = \dfrac{9x - 36 - 6x^2 - 2x}{(2x-5)(3x+1)(x-4)} = \dfrac{-6x^2 + 7x - 36}{(2x-5)(3x+1)(x-4)}$

   d. $\dfrac{3x}{2(x-1)} + \dfrac{7}{4(x+1)} = \dfrac{2(3)x(x+1)}{4(x-1)(x+1)} + \dfrac{7(x-1)}{4(x+1)(x-1)} = \dfrac{6x^2 + 6x + 7x - 7}{4(x-1)(x+1)} =$

   $\dfrac{6x^2 + 13x - 7}{4(x-1)(x+1)}$

13. It appears from the graph that $(2, 0)$ and $(0, 6)$ are points on the line. This would make the slope of the line $\dfrac{6-0}{0-2} = -3$. The equation of the line would be $y = -3x + 6$.

Section 7.5 Review

p. 613

1. a. Slope = 7 b. Vertical Intercept: $(0, 21)$

   c. Horizontal Intercept: $(-3, 0)$

   d. The function is increasing.

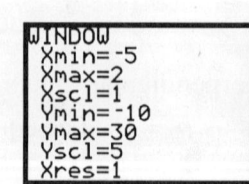

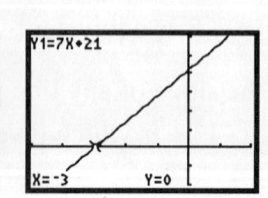

2. a. Slope = $\frac{3}{7}$ b. Vertical Intercept: $(0, 2)$

   c. Horizontal Intercept: $\left(-\frac{14}{3}, 0\right)$

   d. The function is increasing.

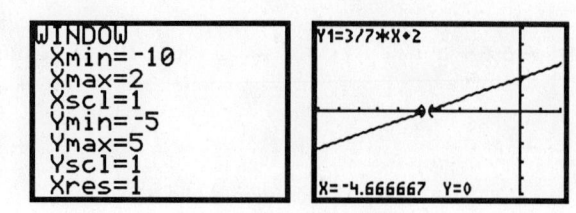

3. a. Slope = 0   b. Vertical Intercept: $(0,0)$

   c. Horizontal Intercept: $(0,0)$

   d. The graph is the *x*-axis and the function is neither increasing nor decreasing.

4. a. Slope = $-\frac{3}{4}$ b. Vertical Intercept: $\left(0, \frac{2}{3}\right)$

   c. Horizontal Intercept: $\left(\frac{8}{9}, 0\right)$

   d. The function is decreasing.

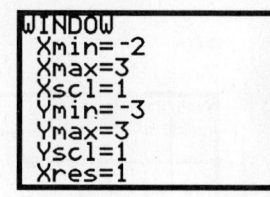

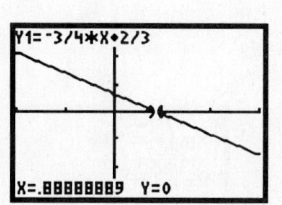

5. a. Slope = $\frac{5}{7}$ b. Vertical Intercept: $\left(0, -2\frac{3}{4}\right)$

   c. Horizontal Intercept: $\left(\frac{77}{20}, 0\right)$

   d. The function is increasing.

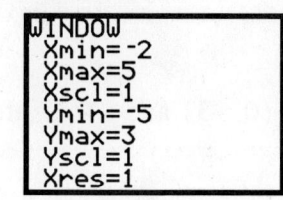

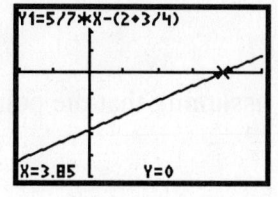

6. a.

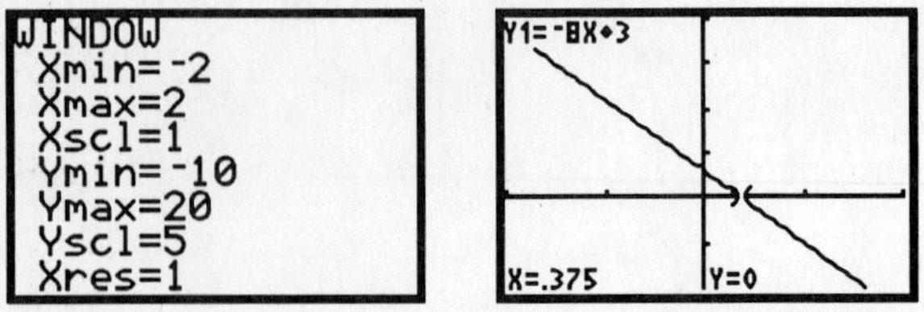

```
WINDOW
 Xmin=-2
 Xmax=2
 Xscl=1
 Ymin=-10
 Ymax=20
 Yscl=5
 Xres=1
```

Y1=-8X+3

X=.375   Y=0

b. $y = -8x + 3$

c. Input at the horizontal intercept: $\dfrac{3}{8}$

d. $y = -8x$

e. $y = \dfrac{1}{8}x + 3$

7. a.

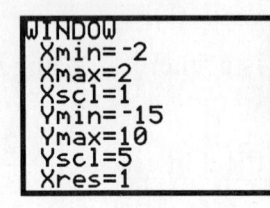

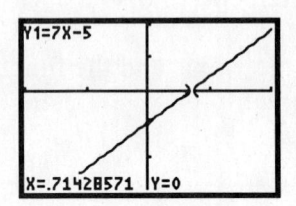

```
WINDOW
 Xmin=-2
 Xmax=2
 Xscl=1
 Ymin=-15
 Ymax=10
 Yscl=5
 Xres=1
```

Y1=7X-5

X=.71428571   Y=0

b. $y = 7x - 5$

c. Input at the horizontal intercept: $\dfrac{5}{7}$

d. $y = 7x - 8$

e. $y = -\dfrac{1}{7}x - 5$

8. a.

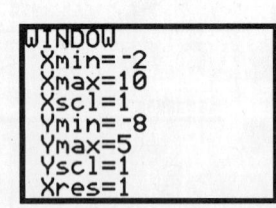

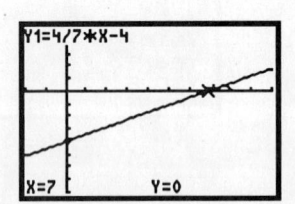

```
WINDOW
 Xmin=-2
 Xmax=10
 Xscl=1
 Ymin=-8
 Ymax=5
 Yscl=1
 Xres=1
```

Y1=4/7*X-4

X=7   Y=0

b. $y = \dfrac{4}{7}x - 4$

c. Input at the horizontal intercept: 7

d. $y = \dfrac{4}{7}x - 7$

e. $y = -\dfrac{7}{4}x - 4$

9. a. $y = x - 3$ assuming that the points $(0, -3)$ and $(4, 1)$ are on the graph.

b. $y = 5x + 7$

156

c. $y = -2x + 3$

d. $y = -4x + 6$ assuming that the points $(0, 6)$ and $(2, -2)$ are on the graph.

10. a. Narrower   b. Narrower and moved down 5 units.

   c. Wider and moved up 4 units.   d. Inverted, narrower and moved up 11 units.

11. a. Zeros: 3 and –3 x-coordinate of the vertex: 0

   b. Zeros: 5 and –4 x-coordinate of the vertex: $\dfrac{1}{2}$

   c. Zeros: –7 and –6 x-coordinate of the vertex: $-\dfrac{13}{2}$

   d. Zeros: 8 and 2 x-coordinate of the vertex: 5

12. a. $y = (x + 8)(x - 7) = x^2 + x - 56$

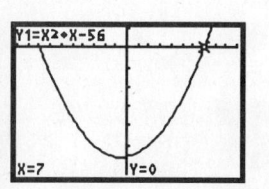

   b. $y = (x + 9)(x + 4) = x^2 + 13x + 36$

c. $y = (x-4)(x+4) = x^2 - 16$

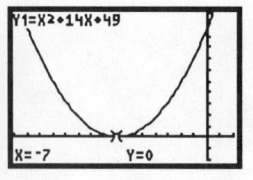

d. $y = (x+7)^2 = x^2 + 14x + 49$

13. a. $(x-10)(x+10)$ b. $(y-9)^2$

    c. prime   d. $2(y-3)(y+8)$

    e. $4(6x+5)$ f. $(t-11)(t-6)$

    g. $5(x+3)^2$ h. $7(x^2+4)$

    i. $(x+8)(x+9)$ j. $4(2x+3)^2$

14. a. $y(-2) = 3(-2) - 5 = -6 - 5 = -11$

    b. $3x - 5 = -2 \quad 3x = 3 \quad x = 1$

15. $0 = -2x - 7 \quad 7 = -2x \quad -\dfrac{7}{2} = x$

16. a. $\begin{aligned} 5x - 7 &= 15 \\ 5x &= 22 \\ x &= \dfrac{22}{5} \end{aligned}$ b. $\begin{aligned} 3 - 11t &= 24 \\ -11t &= 21 \\ t &= -\dfrac{21}{11} \end{aligned}$

c. $4x - 3 > 9$
    $4x > 6$
    $x > \dfrac{3}{2}$

d. $y(y + 18) = -81$
    $y^2 + 18y = -81$
    $y^2 + 18y + 81 = 0$
    $(y + 9)^2 = 0$
    $y = -9$

e. $3 + 2y = 12y - 6$
    $9 = 10y$
    $\dfrac{9}{10} = y$

f. $1 - 2y < 17$
    $-2y < 16$
    $y > -8$

g. $\dfrac{2}{7}x + 1 = \dfrac{3}{4} - 5x$
    $28\left(\dfrac{2}{7}x + 1\right) = 28\left(\dfrac{3}{4} - 5x\right)$
    $8x + 28 = 21 - 140x$
    $148x = -7$
    $x = -\dfrac{7}{148}$

h. $t^2 - 17t + 70 = 4$
    $t^2 - 17t + 66 = 0$
    $(t - 11)(t - 6) = 0$
    $t = 11 \text{ or } t = 6$

i. $3p - 8 = 4 - 5p$
    $8p = 12$
    $p = \dfrac{12}{8} = \dfrac{3}{2}$

j. $2y^2 = 10y + 48$
    $2y^2 - 10y - 48 = 0$
    $2(x - 8)(x + 3) = 0$
    $x = 8 \text{ or } x = -3$

k. $\dfrac{5}{2}t - \dfrac{4}{3} < 9 + \dfrac{8}{6}t$
    $6\left(\dfrac{5}{2}t - \dfrac{4}{3}\right) < 6\left(9 + \dfrac{8}{6}t\right)$
    $15t - 8 < 54 + 8t$
    $7t < 62$
    $t < \dfrac{62}{7}$

l. No value of x will satisfy this inequality.

17. a. The system is equivalent to the system $\begin{array}{l} y = \frac{3}{5}x - \frac{7}{5} \\ y = -4x - 2 \end{array}$. These functions may be graphed in the standard viewing window and the intersect item on the **Calculate** menu can be used to find an approximation to the intersection point.

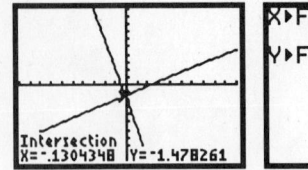

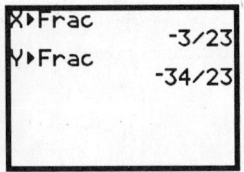

By returning to the home screen and entering X and Y along with the Frac item from the **Math** menu these values can be converted to fraction form. It is difficult to use a table to find the intersection point.

To find the solution by substitution one may solve the equation $\frac{3}{5}x - \frac{7}{5} = -4x - 2$ to get $x = -\frac{3}{23}$ and evaluate one of the functions using this input value to get $y = -\frac{34}{23}$.

b. Solving each equation for y gives us two functions which we can graph. We get $y = -3x + \frac{9}{2}$ and $y = -3x - 1$. We can see that these two functions have graphs which are straight lines with the same slope and different vertical intercepts. The lines are therefore parallel and there is no intersection point. The graphs appear in the standard viewing window.

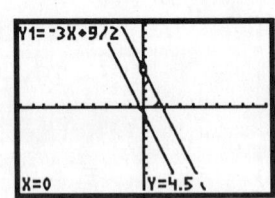

c. We solve each equation for y so that we may enter them into the Y= menu on our TI-83 or 82's. $y = 4x + 8$ and $y = -\frac{3}{2}x + \frac{7}{2}$. The graphs and the intersection point found by using the intersect item on the **Calculate** menu are shown. The standard viewing window has been chosen. The fraction form of the coordinates are also shown.

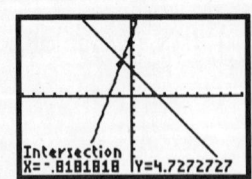

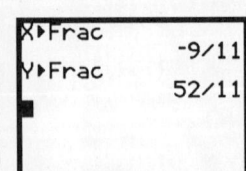

To solve the system by substitution we solve the equation $4x + 8 = -\frac{3}{2}x + \frac{7}{2}$ for x to get

$$2(4x+8) = 2\left(-\frac{3}{2}x+\frac{7}{2}\right) \qquad 8x+16 = -3x+7 \qquad 11x = -9 \qquad x = -\frac{9}{11}.$$

The *y* value is found by evaluating one of the functions:

$$y = 4\left(-\frac{9}{11}\right)+8 = -\frac{36}{11}+\frac{88}{11} = \frac{52}{11}$$

d. The two equations are equivalent to $y = -\frac{1}{5}x+\frac{7}{5}$. Enter these two functions into the
$y = 5x-2$

Y= menu and use the intersect item on the **Calculate** menu to find the intersection point graphically.

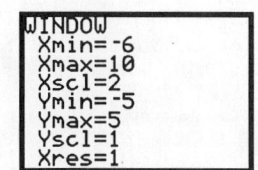

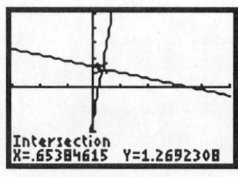

```
WINDOW
Xmin=-6
Xmax=10
Xscl=2
Ymin=-5
Ymax=5
Yscl=1
Xres=1
```

Intersection
X=.65384615 Y=1.2692308

X▶Frac
                17/26
Y▶Frac
                33/26

To find the intersection point by substitution first solve $-\frac{1}{5}x+\frac{7}{5} = 5x-2$ for *x*.

$$5\left(-\frac{1}{5}x+\frac{7}{5}\right) = 5(5x-2)$$

$$-x+7 = 25x-10$$
$$17 = 26x$$
$$\frac{17}{26} = x$$

$$y = 5\left(\frac{17}{26}\right)-2 = \frac{85}{26}-\frac{52}{26} = \frac{33}{26}$$

18. Find the *x*-coordinate by solving $7x+2 = -5x+3$ to get $x = \frac{1}{12}$. The *y*-coordinate is found by evaluating using this value of *x*: $y = 7\left(\frac{1}{12}\right)+2 = \frac{7}{12}+\frac{24}{12} = \frac{31}{12} = 2\frac{7}{12}$.

19.
$$x^2 + 3x = 4$$
$$x^2 + 3x - 4 = 0$$
$$(x+4)(x-1) = 0$$
$$x = -4 \text{ or } x = 1$$

The only solution that makes sense in the problem is the positive one. The width is 1 foot and the length is 4 feet.

20. a. $y = \dfrac{1}{15}x + \dfrac{19}{15}$ b. $y = -\dfrac{13}{4}x + \dfrac{49}{2}$

    c. $y = 3$ d. $x = -5$

21. a. $y = -\dfrac{4}{7}x + \dfrac{9}{7}$ Slope $= -\dfrac{4}{7}$

    Vertical Intercept: $\left(0, \dfrac{9}{7}\right)$ Horizontal Intercept: $\left(\dfrac{9}{4}, 0\right)$

    b. $y = \dfrac{4}{3}x + 5$ Slope $= \dfrac{4}{3}$

    Vertical Intercept: $(0, 5)$ Horizontal Intercept: $\left(-\dfrac{15}{4}, 0\right)$

22. a.

Number of Sweatshirts Sold	Net Profit ($)
0	-500
10	-300
20	-100
30	100
40	300
50	500
60	700

b. $y(x) = 20x - 500$ where x represents the number of sweatshirts sold and $y(x)$ represents the net profit.

c. The ratio of the change in the net profit to the number of sweatshirts sold is $20 per sweatshirt.

d. The graph is shown with the vertical intercept displayed using the Trace feature. The vertical intercept indicates that when they have sold no sweatshirts they will be in the whole $500.

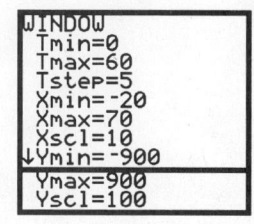

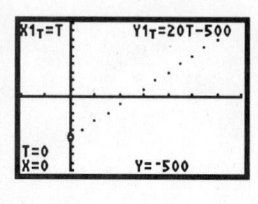

23. a. $\dfrac{12x-8}{3x^2+10x-8} = \dfrac{4(3x-2)}{(3x-2)(x+4)} = \dfrac{4}{x+4}$; $x \neq -4$, $x \neq \dfrac{2}{3}$

b. $\dfrac{9x^2+24x+16}{9x^2-16} = \dfrac{(3x+4)^2}{(3x+4)(3x-4)} = \dfrac{3x+4}{3x-4}$; $x \neq \dfrac{4}{3}$, $x \neq -\dfrac{4}{3}$

24. a. $\dfrac{2}{x-7} - \dfrac{5}{x+4} = \dfrac{2(x+4)}{(x-7)(x+4)} - \dfrac{5(x-7)}{(x-7)(x+4)} = \dfrac{2(x+4)-5(x-7)}{(x-7)(x+4)} =$

$\dfrac{2x+8-5x+35}{(x-7)(x+4)} = \dfrac{-3x+43}{(x-7)(x+4)}$; $x \neq 7$, $x \neq -4$

b. $\dfrac{3x}{2x^2+5x+2} + \dfrac{4}{2x^2+7x+3} = \dfrac{3x(x+3)}{(2x+1)(x+2)(x+3)} + \dfrac{4(x+2)}{(2x+1)(x+2)(x+3)} =$

$\dfrac{3x(x+3)+4(x+2)}{(2x+1)(x+2)(x+3)} = \dfrac{3x^2-9x+4x+8}{(2x+1)(x+2)(x+3)} = \dfrac{3x^2-5x+8}{(2x+1)(x+2)(x+3)}$

$x \neq -\dfrac{1}{2}$, $x \neq -2$, $x \neq -3$

163

TI-82
Graphing Calculator
Reference Manual

TI-82 REFERENCE MANUAL

Turn the calculator on and off

Press **ON** to turn the calculator on. If nothing appears on the screen, adjust the display contrast (see below). If you still see nothing, check the batteries.

Press **2nd** **ON** to turn the calculator off.

Keyboard organization

The keys are grouped by both color and location for ease of use.

The keys are color–coordinated.

To choose a command that is in blue and appearing above a key, press the blue **2nd** key first. By convention, any key that is in blue will appear with black, rather than white, letters in this manual.

To choose a command in gray and appearing above a key, press the gray **ALPHA** key first.

Keys by row beginning at the top of the calculator:

Row 1: Graphing and table keys.

| Y= | WINDOW | ZOOM | TRACE | GRAPH |

Rows 2 and 3: Editing keys.

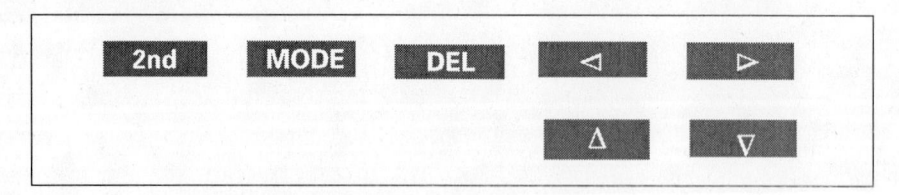

| 2nd | MODE | DEL | ◁ | ▷ |
| | | | △ | ▽ |

Rows 3 and 4: Advanced functions accessible through pull-down menus.

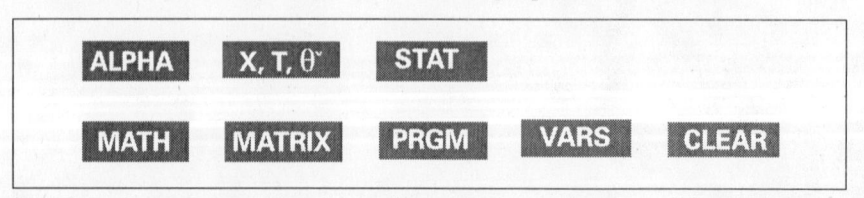

| ALPHA | X, T, θ | STAT |
| MATH | MATRIX | PRGM | VARS | CLEAR |

166

Rows 5–10: Scientific calculator keys.

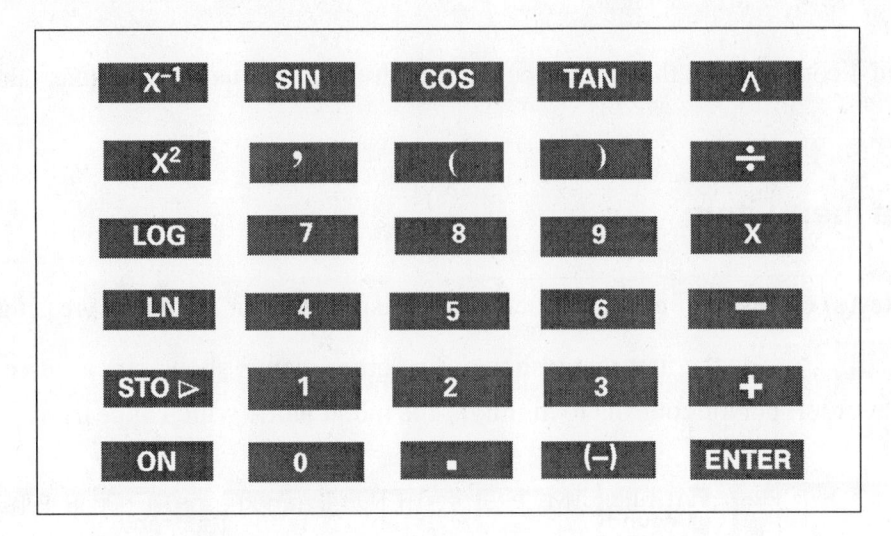

[Calculator keypad image showing keys: X⁻¹, SIN, COS, TAN, ∧; X², ',' , (,), ÷; LOG, 7, 8, 9, ×; LN, 4, 5, 6, −; STO▶, 1, 2, 3, +; ON, 0, ., (−), ENTER]

Adjust the display contrast (darken or lighten screen)

Press and release the **2nd** key.

Press and hold **▲** to increase the contrast (darken the screen).

Press and hold **▼** to decrease the contrast (lighten the screen).

The calculator can be returned to factory settings by pressing **MEM**. Since MEM is in blue above the "+" key, access MEM by pressing **2nd** **+**.

```
MEMORY
1:Check RAM…
2:Delete…
3:Reset…
```

Changing the mode (the Mode Screen)

Press **MODE** to customize the calculator settings. Activated settings are highlighted.

```
Normal Sci Eng
Float  0123456789
Radian Degree
Func Par Pol Seq
Connected Dot
Sequential Simul
Fullscreen Split
```

Change a setting by moving the cursor to the desired setting using the arrow keys.

Press **ENTER** to select a setting.

Menus

Rows 3 and 4 contain keys that allow access to both math menus and variable values.

Selecting a menu item

Press the key(s) corresponding to the menu to be displayed. For example, pressing the **MATH** opens the first menu shown. The other menus shown are accessed by using the right or left cursor controls to highlight the menu name which appears at the top of the screen.

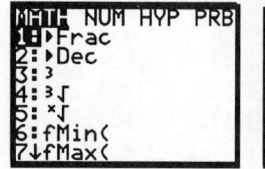

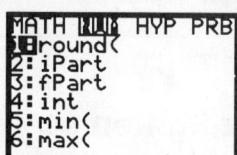

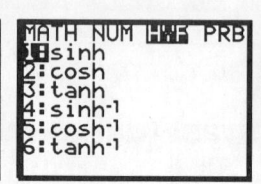

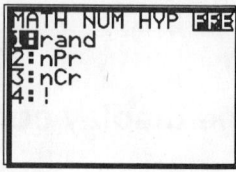

```
MATH NUM HYP PRB      MATH NUM HYP PRB      MATH NUM HYP PRB      MATH NUM HYP PRB
1:▶Frac               1:round(              1:sinh                1:rand
2:▶Dec                2:iPart               2:cosh                2:nPr
3:³                   3:fPart               3:tanh                3:nCr
4:³√                  4:int                 4:sinh⁻¹              4:!
5:ˣ√                  5:min(                5:cosh⁻¹
6:fMin(               6:max(                6:tanh⁻¹
7↓fMax(
```

Select a menu item by either using the arrow keys to highlight the desired choice and then press **ENTER** ; or

if a number is associated with a menu item, press the number next to the menu item.

Home screen

The home screen is used for calculations.

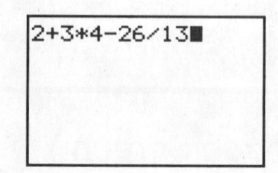

```
2+3*4-26/13■
```

You may return to the home screen at any time by executing the **QUIT** command. This is done by pressing **2nd** **MODE** .

168

Graphics screen

The graphics screen is used to display graphs.

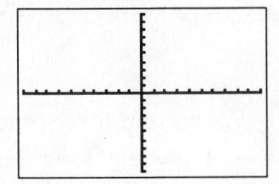

The adjustments to the graphics screen are set using the window screen to be discussed later in the manual. To show the graphics screen press **GRAPH**.

Menu screens

There are several menu screens which display menus of selections which allow you to control the functionality of the calculator and display values of variables.

Table screen

This screen displays a table of inputs and outputs for each active user–defined function.

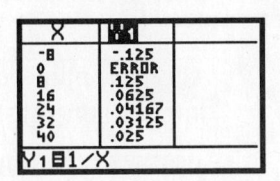

Choose **TABLE** to display this screen by pressing **2nd** **GRAPH**. Notice the word TABLE appears above the GRAPH key printed in blue.

List screen

This screen displays up to six lists where users may enter data.

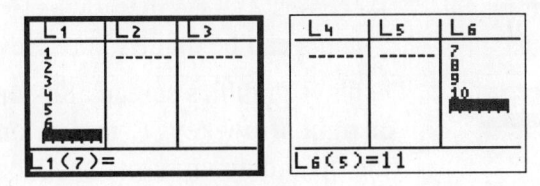

Press **STAT** and then press **1** to choose the Edit item on the menu. This shows the list screen and allows you to edit the list. Press the right cursor control to show more list.

Important keys and their function

Keystroke	Function Description
ON	Turns on calculator; interrupts graph being sketched.
CLEAR	Clears the text screen; deletes functions when in the Y= screen.
ENTER	Executes a command; moves cursor to next line.
(−)	The unary operation of oppositing (not used for subtraction).
MODE	Displays current operating mode.
DEL	Deletes character at the cursor.
^	Symbol used for exponentiation
Y=	Accesses screen where up to eight functions may be entered and stored.
X, T, θ	Enters the variable x in function mode; enters the variable t in parametric mode; enters Θ in polar mode.
MATH	Accesses numerical representations; random numer generator and combinatoric entries.
2nd	Accesses functions and menus printed in blue.
ALPHA	Accesses letters and symbols printed in grey to the right above many keys.
STO ▷	Used to assign a numeric value to a variable
GRAPH	Displays the graphical representation of entered functions.
WINDOW	Accesses menus for setting the viewing window and selecting the window format.
ZOOM	Accesses ZOOM menu where various adjustments can be made to the viewing window.
TRACE	Displays graphics screen and cursor. Use the left or right arrow keys to trace along a graph. Use the up and down arrow keys to move the trace cursor from one function to another function.

Accessing functions using the [2nd] key
(commands written in blue, above and left of center of the keys).

Editing Functions Accessed with the [2nd] key.

Function	Keystrokes	Description
OFF	2nd ON	Turns the calculator off.
ENTRY	2nd ENTER	Allows the last line entered to be edited and re–executed.
INS	2nd DEL	Allows insertion of character(s) to the left of the cursor.
LINK	2nd STAT	Allows transfer of data or programs between two TI-82 calculators.
QUIT	2nd MODE	Exits menus; returns calculator to home screen.

Table Functions Accessed with the [2nd] key.

Function	Keystrokes	Description
TABLE	2nd GRAPH	Displays a table of x–values and corresponding output values functions stored in the **Y=** menu.
TblSet	2nd WINDOW	Accesses the TABLE SETUP menu where the minimum table value and the table increment are set.
STATPLOT	2nd Y=	Accesses the STAT PLOT menu where statistical data may be graphically displayed using a histogram, point plot, line plot, or box–and–whisker plot.

Calculations

Calculations are done in the home screen. If you are not in the home screen, choose `QUIT` which is accessed by pressing `2nd` `MODE`.

Enter expressions as you would write them. Observe the algebraic order of operations.

After you complete typing an expression, press `ENTER` to obtain the answer.

If you make a mistake during entry, use the arrow keys, `INS` (for insert), or `DEL` (for delete) to edit your expressions.

Raising a number to a power `^`

Enter an exponent first entering the base and then pressing the `^` key followed by the exponent. For example, 4^5 appear on the home screen as shown. Press `ENTER` to display the value of the expression.

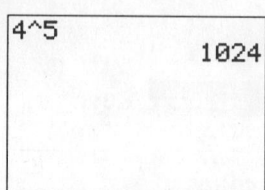

```
4^5
        1024
```

Negative numbers `(-)`

Enter a negative number using the *gray* key `(-)` (located to the left of the `ENTER` key), not the blue key, which is reserved for the operation of subtraction.

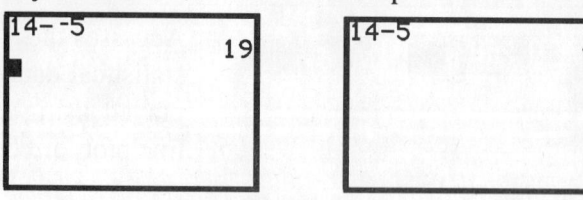

```
14--5
         19
```

```
14-5
          9
```

Scientific notation (3.46E11 means 3.46×10^{11})

This notation is uses powers of ten to represent very large and very small numbers. Numbers are written as a decimal times a power of 10.

To enter on the calculator, replace the **X 10** with **EE**. The exponent appears immediately to the right of the E in the display. The display, 3.46E11, actually represents 3.46×10^{11}.

Enter **EE** by pressing [2nd] [,] . Only one E appears on the screen to indicate that the next number entered will be accepted as a power of ten.

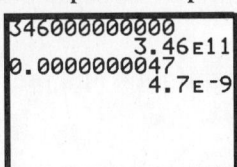

```
346000000000
         3.46E11
0.0000000047
         4.7E-9
```

Editing a previously-entered expression [2nd] [ENTER]

The last entered in the calculator can be displayed on the screen and edited. Execute the **Entry** command to display the last expression entered by pressing the following keys.

Choose [ENTRY] by pressing [2nd] [ENTER].

A copy of the previously entered expression should appear on the screen. This expression may be edited. The new expression may be executed by pressing [ENTER].

Reciprocal function [x⁻¹]

To find the reciprocal of a number first enter the number whose reciprocal you wish to find on the home screen. If the number is a fraction it must be enclosed in parentheses.

Press [x⁻¹].

Press [ENTER] to display the reciprocal of the original number in decimal form. To display the answer as a fraction choose item number one from the math menu by pressing [MATH] then [1].

```
(7/3)⁻¹
         .4285714286
Ans►Frac
               3/7
```

173

INT function

The greatest integer function (**int**) returns the largest integer less than or equal to a number, expression, list or matrix. For nonnegative numbers and negative integers, the **int** value is the same as **iPart**. For negative noninteger numbers **int** returns a value one integer less than **iPart**.

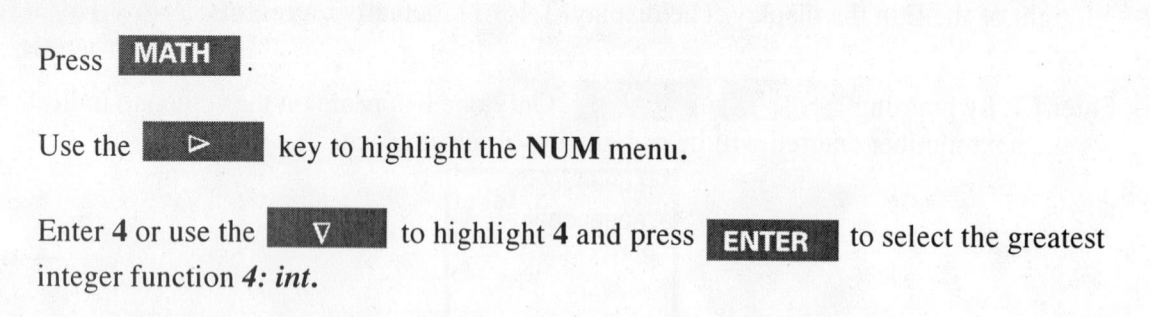

Press **MATH**.

Use the **▷** key to highlight the **NUM** menu.

Enter **4** or use the **▽** to highlight **4** and press **ENTER** to select the greatest integer function *4: int*.

The calculator returns to previous screen and inserts **int** function at cursor. Follow the function name with an input value in parentheses. The function returns an output which is the largest integer that does not exceed the input.

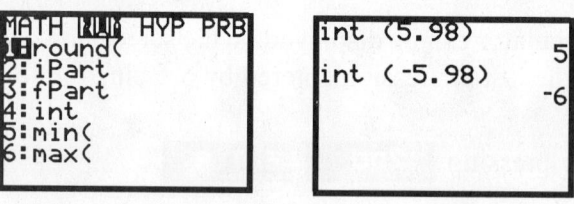

Obtaining fractional results

If a computation results in a decimal output, the result can often be converted to a fraction by using the **Frac** command.

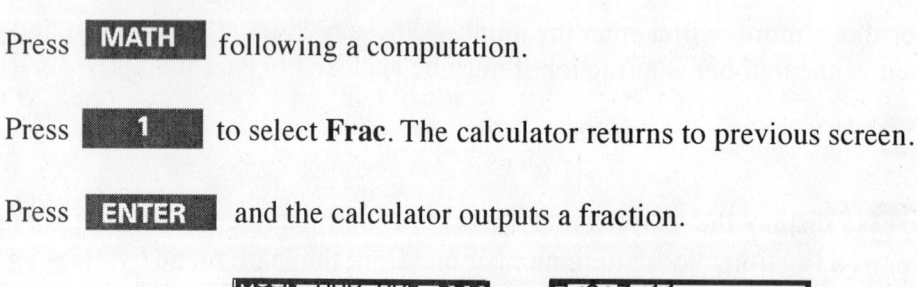

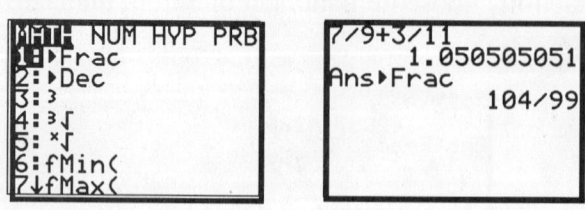

Press **MATH** following a computation.

Press **1** to select **Frac**. The calculator returns to previous screen.

Press **ENTER** and the calculator outputs a fraction.

RAND function

The calculator contains a random number generator that will return random numbers greater than zero and less than one.

The random number generator should be seeded initially so that each calculator will generate a different sequence of numbers. To control a random number sequence, first store an integer seed value in **rand**. The factory-set seed value is used whenever you reset the TI-82 or when you store **0** to **rand**.

Start in the home screen.

Type a favorite whole number. This number acts as the seed.

Press **STO ▷**. This is the key used for storing a value in the calculator.

Press **MATH**.

Use the **◁** key to highlight the **PRB** menu.

Press **1** to select **rand**.
The calculator returns to the home screen and inserts the **rand** function.

Press **ENTER** to complete the seeding process.

Recall **rand** by pressing **MATH**, arrowing to **PRB**, and selecting **1**.

Press **ENTER** to see the first random number.

Continually pressing **ENTER** will generate additional random numbers.

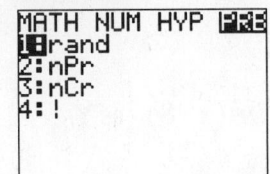

```
MATH NUM HYP PRB
1:rand
2:nPr
3:nCr
4:!
```

```
83→rand
                83
rand
        .8815441117
        .0397447541
        .7048789638
        .9460171244
```

Factorial function

The factorial of a whole number is the product of all whole numbers less than or equal to the given whole number.

   Type a whole number.

   Press **MATH**.

   Use the **◁** key to highlight the **PRB** menu.

   Press **4** to select **!** which represents factorial.

The calculator returns to the home screen and inserts the **!** after the number you typed.

   Press **ENTER** to see the result of the factorial.

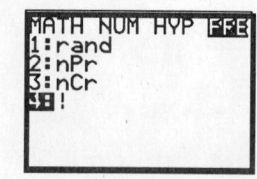

```
MATH NUM HYP PRB
1:rand
2:nPr
3:nCr
4:!
```

```
13!
       6227020800
```

Entering expressions in the Y = menu **Y=**

When you enter a function the = symbol is highlighted to show the function is selected to be graphed. The highlighting may be toggled on and off by using the arrow keys to move the cursor to the = symbol and then pressing **ENTER**.

Define the function:

   Press **Y=** to access the function entry screen.

   Type the expression for the function after $Y_1 =$. This defines the function as Y_1.

Select the *int* function:

   Example: Enter the function $f(x) = x - 8int(x/8)$ as Y_1:

   Press **Y=** to access the function entry screen.

   Enter the function expression $x - 8$ at the cursor.

   Press **MATH**.

Use the [▷] key to highlight the **NUM** menu.

Press the **4** key or use the [▽] to highlight **4** and press [ENTER] to select the greatest integer function *4: int*.

Complete the entry of the function by entering **(x / 8)** then press [ENTER].

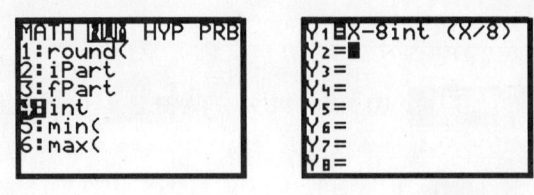

Functions may be removed from this menu by pressing [CLEAR].

Up to ten functions can be defined and accessed simultaneously.

Displaying a table of values from a function in the Y= menu

The TI–82 will display a table of values for a function defined under the Y = menu.

Example: Display a table of values for the function $y(x) = x - 8int\left(\frac{x}{8}\right)$

First Define the function:

Press [Y=] to access the function entry screen and enter *x - 8 int (x/8)* as Y_1.

Press [2nd] [WINDOW] to access [TblSet], the table set up menu.

Set the beginning table value and the increment:

Enter a beginning table value for **TblMin** at the flashing cursor.

Press the [▽] to place the cursor on the number next to Δ**Tbl =** to change the increment value.

Δ**Tbl** represents the increment on the input values. The default value is **1**. If you wish a different increment, enter that value.

Check to see that **Auto** is highlighted for both Indpnt and Depend.

If not, use arrow keys to place cursor on **Auto** and press `ENTER` to select it.

```
TABLE SETUP
 TblMin=-3
 ΔTbl=1
 Indpnt: Auto Ask
 Depend: Auto Ask
```

To view the table:

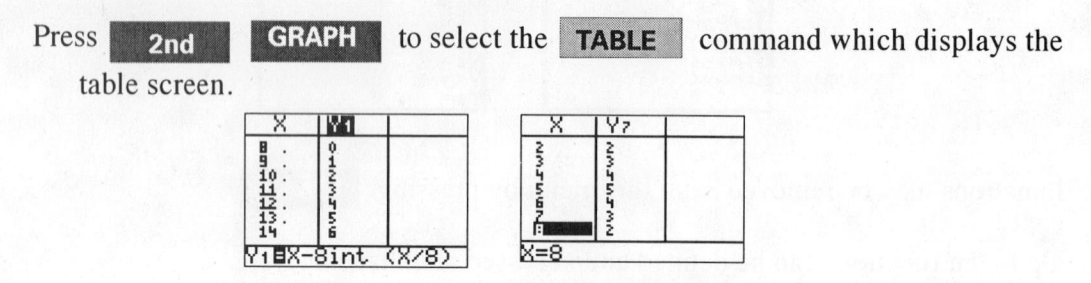

Press `2nd` `GRAPH` to select the `TABLE` command which displays the table screen.

By using the up and down arrows, you can scroll through the table, either up or down.

Storing a value for a variable

In the home screen enter the input value and store it in the variable x.

Type the number you wish to store.

Press `STO ▷`. This is the key used to store values.

When pressed, an arrow pointing right appears.

Press `X, T, Θ`. This will display the variable x.

```
3→X
```

Press `ENTER`. this causes the storage to take place.

A common mistake is to forget to press `ENTER`.

Example: Store the value of 3 in x.

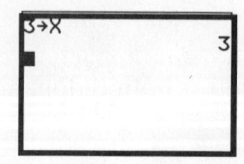

```
3→X        3
```

Function evaluation: storing input

Example: Evaluate the function $y(x) = -x^2 + 8x + 19$ for $x = 3$.

Once a function is defined using the **Y =** menu, you can evaluate the function for a given input value by storing the value in the variable x.

Define the function:

Press [Y=] to access the function entry screen.

Type the expression $-x^2 + 8x + 19$ for the function after $Y_1 =$.
This defines the function as Y_1.

Return to the Home Screen:

Return to the Home Screen by pressing [2nd] [MODE].

Enter the **input value** followed by [STO ▷] [X, T, Θ].
This will have the effect of storing a value for variable x.

Writing more than one command on the same line (chaining commands):

Enter a colon by pressing [2nd] [.]. The colon is used to chain two or more commands together on the same line.

Select the function you wish evaluated:

Select Y_1 by pressing [2nd] [VARS] [1] [1].

Evaluate the function:

Press [ENTER]. The function value is displayed.

```
Y1∎-X²+8X+19         3→X:Y1
Y2=                         34
Y3=
Y4=
Y5=
Y6=
Y7=
Y8=
```

Function evaluation using function notation

To evaluate a function using function notation you must first define the function in the Y = menu as one of the Y variables. Then on the Home Screen enter the Y variable with the input value in parentheses next to that Y variable.

Example: Evaluate the function $y(x) = -x^2 + 8x + 19$ at $x = 3$.

First, define the function:

Press **Y=** to access the function entry screen.

Type the expression $-x^2 + 8x + 19$ for the function after $Y_1 =$. This assumes that Y_1 is clear. This defines the function as Y_1.

Second, return to the Home Screen:

Return to the home screen by pressing **2nd** **MODE**.

Third, select the Y variable you wish evaluated:

Type **2nd** **VARS** **1** **1** to access Y_1 in the Y variables menu. Y_1 appears in the home screen.

Finally, complete the process by giving the input value in parentheses:

Immediately after Y_1,

Type (3) followed by **ENTER**.

This returns the output of the function stored in Y_1 at an input value of 3.

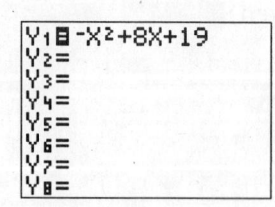

```
Y1■-X²+8X+19
Y2=
Y3=
Y4=
Y5=
Y6=
Y7=
Y8=
```

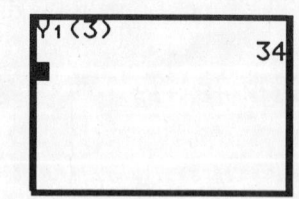

```
Y1(3)
              34
```

Function evaluation using Ask mode with the table screen

You can define a function and, using the table with the independent variable set to Ask mode and the dependent variable set to Auto mode, enter values for the independent variable and by pressing ENTER after each entry, displaying the value of the function for the input entered.

Evaluate the function $y(x) = -x^2 + 8x + 19$ at $x = 3$.

Define the function:

Press **Y=** to access the function entry screen.

Type the expression $-x^2 + 8x + 19$ for the function after $Y_1 =$. This assumes Y_1 is clear. This defines the function as Y_1.

Select the table setup **TblSet** :

Press **2nd** **WINDOW** to access **TblSet**, the table set up screen.

Arrow down to **Indpnt** and over to **Ask**.

Press **ENTER** to select the **Ask** mode for the independent variable. The dependent variable should still be in **Auto** mode.

Press **2nd** **GRAPH** to access **TABLE**, and display the table screen. A blank table appears.

Enter the input value:

Type the input followed by **ENTER**. The output for this input appears.

You may continue to enter inputs and the calculator will return the corresponding outputs.

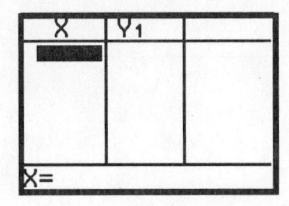

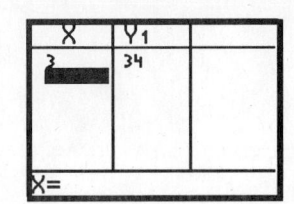

Graphing a function

Once a function is defined in the Y = menu, its graph can be displayed on the graphics screen. It will be necessary to adjust the viewing rectangle which controls the portion of the graph shown.

Example: Graph the function $f(x) = -x^2 + 8x + 19$.

Define the function:

Press **Y =** to access the function entry screen.

Type the expression $-x^2 + 8x + 19$ for the function after $Y_1 =$.
This defines the function as Y_1.

Choose a viewing window:

A *viewing window* is a rectangular portion of the coordinate plane.

Xmin puts the left edge of the viewing window at this x-coordinate.

Xmax puts the right edge of the viewing window at this x-coordinate.

Xscl defines the distance between tick marks on the horizontal axis.

Ymin puts the lower edge of the viewing window at this y-coordinate.

Ymax puts the upper edge of the viewing window to this y-coordinate.

Yscl defines the distance between tick marks on the vertical axis.

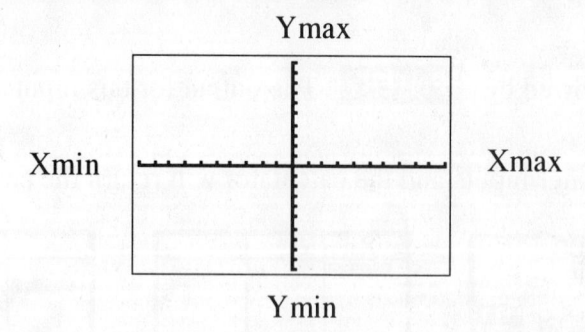

Standard viewing window

The *standard viewing window* is defined as:

Xmin = –10	Ymin = –10
Xmax = 10	Ymax = 10
Xscl = 1	Yscl = 1

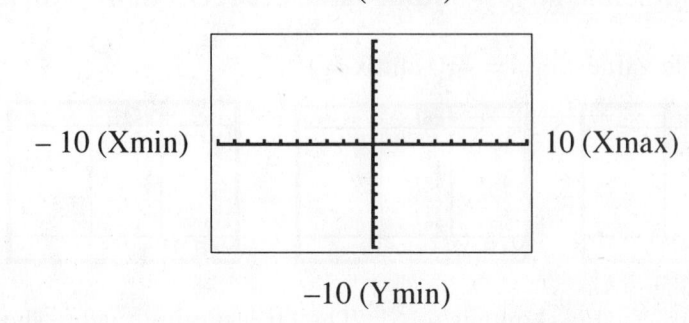

10 (Ymax)

– 10 (Xmin)   10 (Xmax)

–10 (Ymin)

Set the view window:

Press WINDOW and enter the viewing window values.

Press the GRAPH key to display the graphics screen with a portion of the graph shown.

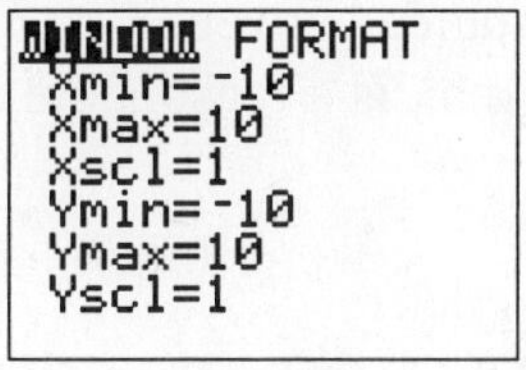

```
WINDOW FORMAT
Xmin=-10
Xmax=10
Xscl=1
Ymin=-10
Ymax=10
Yscl=1
```

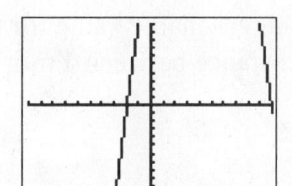

To select the standard viewing window using the Zoom menu:

Press ZOOM 6.

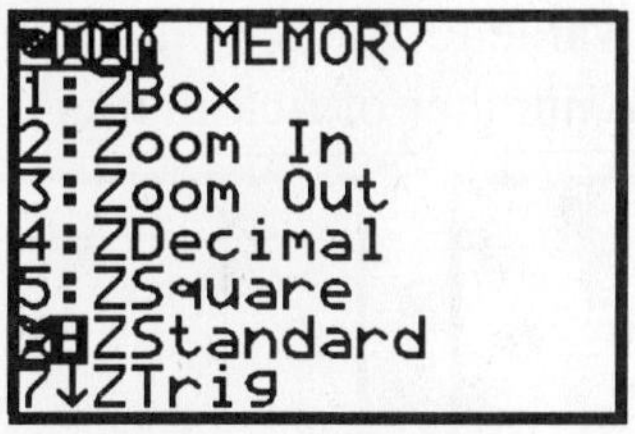

```
ZOOM MEMORY
1:ZBox
2:Zoom In
3:Zoom Out
4:ZDecimal
5:ZSquare
6:ZStandard
7↓ZTrig
```

Changing the viewing window

The previous graph is only a portion of the complete graph. To see more of the graph, change the viewing window. This is done by pressing the **WINDOW** key which displays the window screen. First decide how to change the window. The table feature helps determine a better viewing window.

Press **2nd** **GRAPH** to display the table.

Determine the smallest and largest output value between Xmin (–10) and Xmax (10).

Examine the table values for x = –10 and x = 10.

The smallest output value shown is –161. The largest output value shown is 35.

Press **WINDOW**.

Enter a value for Ymin which is smaller than the smallest output –161. –170 would be a good choice.

Enter a value for Ymax which is larger than the largest output value 35. Say, 40.

For Yscl, choose a reasonable value for the distance between tick marks. Approximately one-tenth of the distance between Ymin and Ymax is a reasonable choice. Set Yscl = 20.

Graph the function:

Press **GRAPH**. A more complete graph of the function appears.

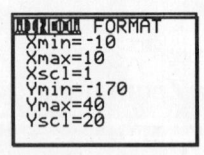

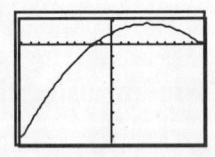

Change Xmin and Xmax to see more of the graph to the left or to the right. For instance one might increase the Xmax value to 20. It would then be wise to increase the Xscl value so there won't be an excessive number of tick marks on the x-axis.

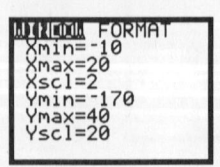

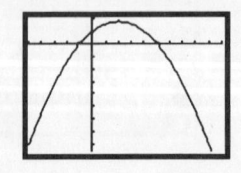

Trace: displaying coordinates of points on a graph

The cursor can be moved from one plotted point to the next along a defined function. The coordinates of the point located by the cursor are displayed at the bottom of the screen.

Example: Trace along the graph of $f(x) = -x^2 + 8x + 19$.

Define the function:

Enter the expression $f(x) = -x^2 + 8x + 19$ as Y_1 at the Y = menu.

Verify the viewing window settings:

If the viewing window is not in standard setting, enter the following:

Xmin = –10	Ymin = –170
Xmax = 10	Ymax = 40
Xscl = 1	Yscl = 20

Graph the function:

Press GRAPH . The graph of the function is displayed.

Press TRACE .

A flashing cursor appears on the graph along with a value for input and output at the bottom of the screen.

Press the left arrow key to move the cursor toward smaller inputs. Moving to the left of Xmin causes the graph to scroll by automatically decreasing Xmin.

Press the right arrow key to move the cursor toward larger inputs. Moving to the right of Xmax causes the graph to scroll by automatically increasing Xmax.

As the cursor moves, the values of the x and y-coordinates at the cursor appear at the bottom of the screen.

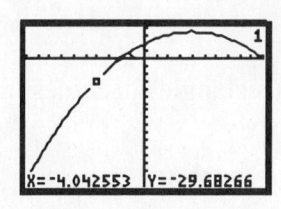

Note that the choices for the x-coordinates are determined by the programing of the graphing utility. The coordinates which are displayed can be controlled. as is explained in the next section of this manual.

Friendly viewing windows to control the Trace Display

A "friendly" viewing window displays values for x with at most one decimal position. There are 95 pixels across the screen. The distance between Xmin and Xmax is divided by 94 to determine the size of the increment in x when moving the cursor in the graphics window.

To obtain a "friendly" window, Xmin and Xmax are selected so that when the difference Xmax − Xmin is divided by 94, the quotient terminates by the tenths position.

Tracing by x increments of 0.1: Zoom Menu Item Number 4:ZDecimal

Press **ZOOM** **4** to automatically set the "friendly" viewing window to

Xmin = −4.7   Ymin = −3.1

Xmax = 4.7   Ymax = 3.1

Xscl = 1   Yscl = 10

Notice that Xmax − Xmin = 9.4. Divide this difference by 94, to determine the x-increment displayed. The quotient is 0.1. The x values displayed by the Trace feature will increment by 0.1 in this window.

Using the ZOOM options to adjust the viewing window

Use the **ZOOM** key to open the following menu of options for altering the viewing window.

1:ZBox: creates a new viewing window based on a rectangle that you draw.

Press **ZOOM** **1** to return to the graphics window.

Use the arrow keys to move the cursor to one corner of the desired viewing window.

Press **ENTER** to set one corner.

Move the cursor to the opposite corner of the desired viewing window.

Press **ENTER** and the box (rectangle) becomes the new viewing window.

2:Zoom In: Use this item to zoom in using a preset factor to decrease the size of the viewing window. The factors can be set by selecting the Memory submenu from the Zoom screen and selecting item number **4:SetFactors** in that submenu.

Press **ZOOM** **2** which returns you to the graphics window.

Use the arrow keys to position the cursor at the desired center of the new window.

Press **ENTER** .

You can now repeatedly zoom in by moving the cursor to the center of the desired window and pressing **ENTER** .

3:Zoom Out: Use this item to zoom out using a preset factor to increase the size of the viewing window. The factors can be set by selecting the Memory submenu from the Zoom screen and selecting item number **4:SetFactors** in that submenu.

Press **ZOOM** **3** which returns you to the graphics window.

Use arrow keys to position the cursor at the desired center of the new window.

Press **ENTER** .

You can now repeatedly zoom out by moving the cursor to the center of the desired window and pressing **ENTER** .

4:Decimal: This item sets a viewing window so that the x values will increment by 0.1 as the cursor moves across the window or when using the trace feature. These are "friendly" values for analyzing the graph.

Press **ZOOM** **4** to change the window so that Xmin = −4.7, Xmax = 4.7, Ymin = −3.1, and Ymax = 3.1. Both Xscl and Yscl are 1.

5:Square: sets the viewing window so that the aspect ratio is 1. Graphs, such as circles, will appear distorted if the aspect ratio is not one.

Press **ZOOM** **5** .

6:Standard: returns graph to the standard viewing window.

Press **ZOOM** **6** to graph in the standard viewing window.

7:Trig: sets a "friendly" viewing window for graphing trigonometric functions.

Press **ZOOM** **7** to change the window.

8:Integer: sets the viewing window so that the *x* values will increment by one as the cursor moves across the window. These are "friendly" values for analyzing the graph.

Press **ZOOM** **8** to change the window.

9:ZoomStat: This item resets the viewing window so that all the points in the active StatPlots are displayed. Press **ZOOM** **9** to change the window.

Connected and Dot Modes for graphs

By choosing the MODE setting on the fifth line to the MODE screen, you can display either a connected graph, where the points plotted are connected to make the graph look relatively smooth, or a discrete graph of a function, where only the plotted points are shown. The calculator comes from the factory set in **Connected** mode. This is the default setting. This setting must be changed to **Dot** to view a discrete graph.

As an example the graph of the function $y(x) = 2x - 1$ is shown with both mode settings using the "friendly" viewing window set by choosing item number 4:ZDecimal in the Zoom menu.

Dot mode

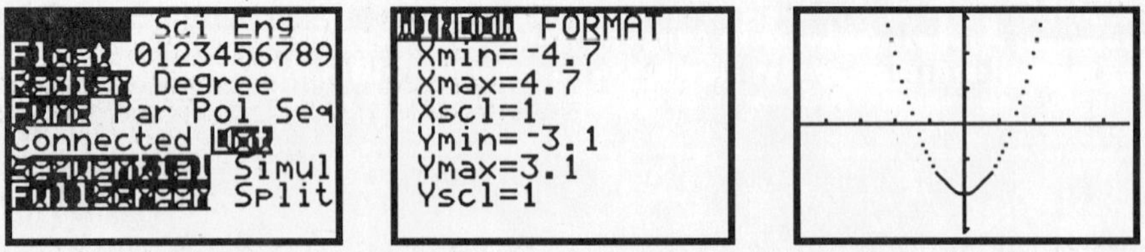

Connected mode

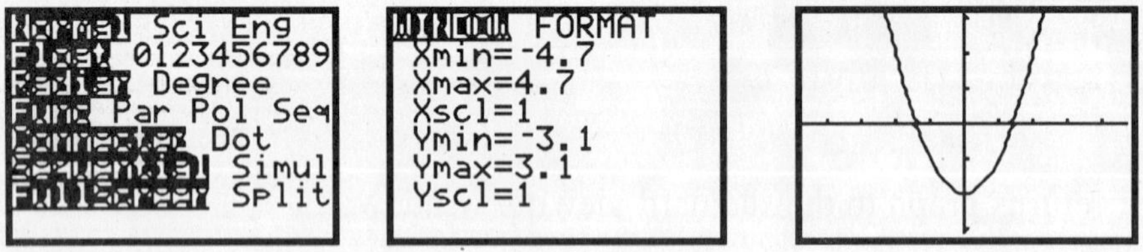

188

Parametric graphing

Parametric equations consist of two equations, representing the *x*-component and the *y*-component, each expressed in terms of a third independent variable, *T*. Parametric equation may be used to control the points which are plotted on the graph of a function.

Before entering the two parametric equations, you must change from function mode to parametric mode and adjust your viewing window.

Press **MODE** and use the **▽** key to highlight **Par**.

Press **ENTER** .

Highlight **Dot**.

Press **ENTER** . The mode is now set appropriately for parametric plotting.

Example: Suppose we wish to plot the function $y(x) = 10x - 1000$ but only show the points on the graph whose x-coordinates are 0, 5, 10,... 200. First we select Parametric and Dot in the Mode window.

Now we open the Y= menu and define $x = t$ and $y = 10t - 1000$.

Press **Y=** and enter *t* for X_{1T} and $10t - 1000$ for Y_{1T}.

Press **WINDOW** and set the viewing window appropriately. (Shown below.)

The value of Tmin will determine the smallest *t* and hence *x*- coordinate to be plotted and the Tmax will determine the largest value of t and therefore *x* to be plotted. The size of the Tstep will determine the distance between the *x*-coordinates of the points which are plotted. The larger Tstep, the fewer points will be plotted.

Press **GRAPH** when all this is set up.

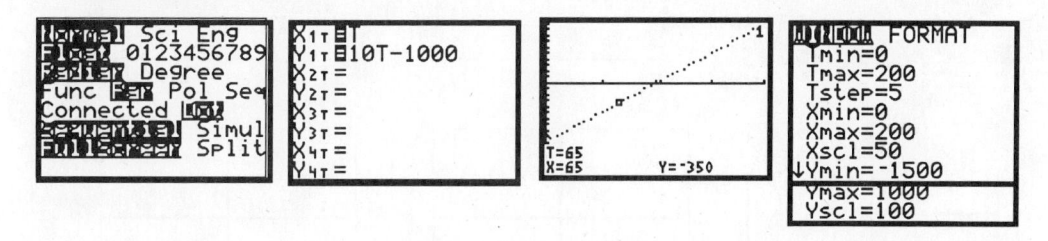

When you use the trace feature you will see the points on the graph highlighted by the cursor and the *t*, *x* and *y* coordinates will be displayed at the bottom of the screen.

Graphing the solution interval to inequalities in one variable

Inequalities in one variable, such as $2x + 3 < 5$, can be entered directly in the Y= menu. The calculator will graph the solution set by drawing a horizontal line at $y = 1$ for all x values that make the inequality true. The "1" is the calculator's way of saying "true" whereas "0" is what the calculator uses for "false".

Set the calculator to **Dot** mode:

   Press MODE , arrow down and over to **Dot**, and press ENTER .

   Press Y= .

   Type the left hand side of the inequality.

   Open the **TEST** menu by pressing 2nd   MATH .

Select the desired inequality relation by typing the number next to the relation, in this case **5**. The calculator returns to the Y= screen.

   Type the right hand side of the inequality.

Set an appropriate viewing window. It is important that $y = 0$ and $y = 1$ be clearly visible on the graphics screen. It would be appropriate to use $Y\min = -2$ and $Y\max = 2$.

   Press GRAPH . A horizontal line segment at $y = 1$ appears for x values that satisfy the inequality.

You can trace on the graph. An output of 1 indicates the corresponding x is in the solution set to the inequality. An output of 0 indicates the corresponding x makes the inequality false.

   Example: Graph the solution set to $2x + 3 < 5$.

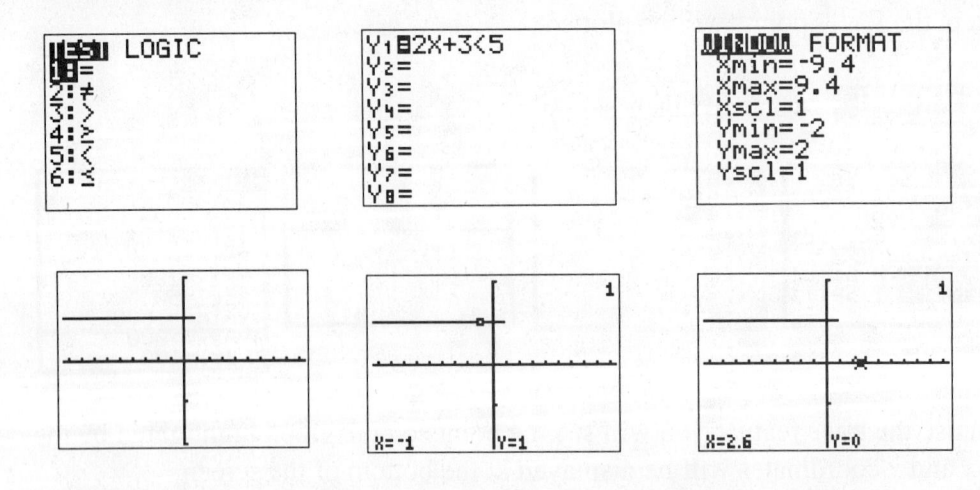

Finding zeros using the CALCULATE menu item 2:root

Enter the function in the Y= menu(the mode setting should be **Func**) and graph the function in a window so that the points where the graph crosses the horizontal axis are visible in the graphics screen.

Press **2nd** **TRACE** to display the **CALCULATE** menu, abbreviated as CALC in blue above the Trace key.

Type **2** to select **2:root** from the menu.

The calculator returns to the graphics window and prompts you to select a **Lower Bound** for one of the roots.

Use the **left arrow key** to move the cursor until the value for x at the cursor is smaller than the desired zero or root.

Press **ENTER** and the x value of the cursor is selected as the lower bound.

Now the calculator prompts you to select an **Upper Bound** for the root.

Press the **right arrow key** at least twice to move the cursor until the value for x at the cursor is larger than the desired zero.

Press **ENTER** and the x value of the cursor is selected as the upper bound.

The calculator prompts you to select a **Guess** for the root.

Press the **left arrow key** to move the cursor close to the zero which is where the graph intersects the x-axis. Be careful not to move the cursor past the lower bound. The guess must be between the lower and upper bounds. When the cursor is close to the root, press **ENTER**.

The calculator begins approximating the zero. A moving vertical line segment is displayed in the upper right corner of the screen to indicate that the calculator is computing. When finished, the calculator displays the approximate zero within the interval defined by the lower and upper bounds. This solution is correct with an error of no more than 0.00001.

Example: Find the largest zero of the function $y = x^2 + x - 3$.

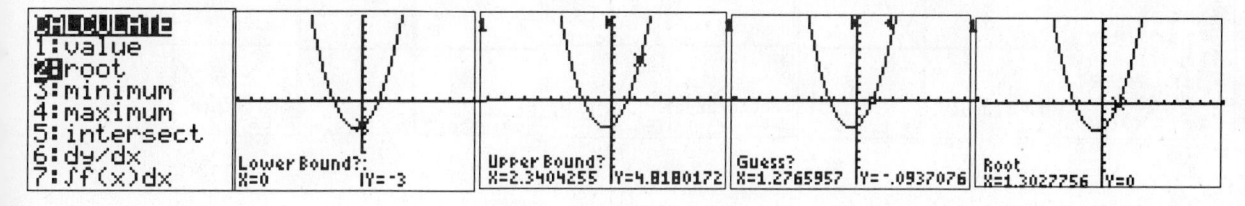

The largest zero or root of the function is 1.3027756, approximately.

Finding intersection points using the CALCULATE menu

Enter the two functions in the Y= menu as Y_1 and Y_2. (You may use any of the other y variables if you do not wish to erase the functions you have entered for Y_1 and Y_2.)

Graph both functions in a window so that the desired intersection point is visible on the graphics screen.

Press **2nd** **TRACE** to display the **CALCULATE** menu.

Type **5** to select **5:intersect.**

The calculator returns to the graphics screen. The calculator prompts you to select the **First curve**, meaning the graph of one of the functions. The cursor is on the graph of Y_1.

Press **ENTER** to select Y_1 as the first curve.

Now the calculator prompts you to select the **Second curve**. The cursor is on the graph of Y_2. (If there were more than two graphs on the graphics screen you could move the cursor to one of the other graphs by using the up or down cursor control.)

Press **ENTER** to select Y_2 as the second curve.

The calculator prompts you to supply a **Guess** for the intersection point.

Do this by using the arrow keys to move the cursor near the intersection point.

Press **ENTER** when the cursor is close to the intersection point.

The calculator begins approximating the intersection point. A moving vertical line segment is displayed in the upper right corner of the screen to indicate that he calculator is computing. When finished, the calculator displays the approximate ordered pair for the intersection point. The solution is correct with an error of no more than 0.00001.

Example: Find the intersection point of the functions $y = 2x - 1$ and $y = -3x + 7$. The graph appears in the standard viewing window.

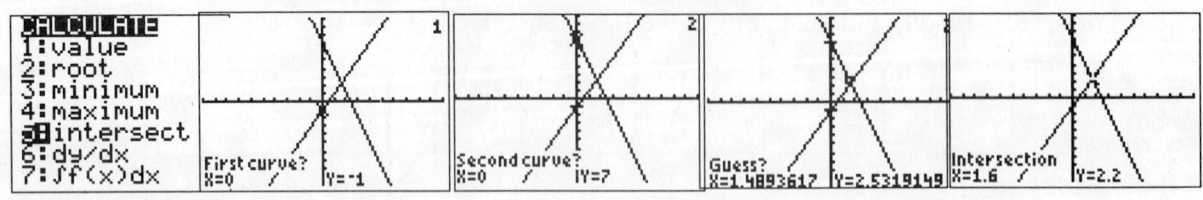

The intersection point is (1.6, 2.2).

Finding a function's maximum using the CALCULATE menu

Enter the function in the Y= menu.

Graph the function in a window so that the graph displays a highest point.

Press **2nd** **TRACE** to open the **CALCULATE** menu.

Type **4** to select **4:maximum**.

The calculator returns to the graphics window. The calculator prompts the user to select a **Lower Bound** for the interval containing the function's maximum.

> Use the **left arrow key.** Move the cursor until it is to the left of the function's maximum.
>
> Press **ENTER** to select this x value as the lower bound.

Now the calculator prompts the user to select an **Upper Bound** for the interval containing the function's maximum.

> Press the **right arrow key** at least twice to move the cursor until the cursor is to the right of the function's maximum.
>
> Press **ENTER** to select this x value as the upper bound.

The calculator prompts the user to select a **Guess** for the function's maximum.

> Press the **left arrow key** to move the cursor close to the function's maximum.

Be careful not to move the cursor past the lower bound. The guess must be between the lower and upper bounds.

> Press **ENTER** when the cursor is close to the maximum.

The calculator begins approximating the function's maximum. A moving vertical line segment is displayed in the upper right corner of the screen to indicate that the calculator is computing. When finished, the calculator displays the approximate maximum of the function within the interval defined by the lower and upper bounds. This solution is correct with an error of no more than 0.00001.

> *Example:* Find the vertex of $y = -x^2 + x + 4$. For this function the vertex is the highest point on the graph. The graph appears in the standard viewing window.

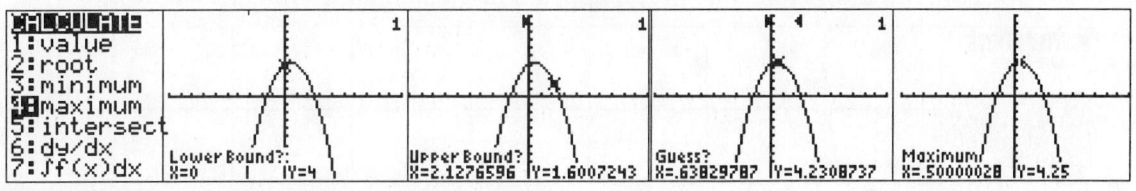

```
CALCULATE
1:value
2:root
3:minimum
4:maximum
5:intersect
6:dy/dx
7:∫f(x)dx
```
Lower Bound? X=0   Y=4
Upper Bound? X=2.1276596   Y=1.6007243
Guess? X=.63829787   Y=4.2308737
Maximum X=.50000028   Y=4.25

The vertex is actually the point (0.5, 4.25).

193

Finding a function's minimum using CALC

Enter the function in the Y= menu.

Graph the function in a window so that the graph displays a lowest point.

Press **2nd** **TRACE** to open the **CALCULATE** menu.

Type **3** to select **3:minimum**.

The calculator returns to the graphics window. The calculator waits for you to select a **Lower Bound** for the interval containing the function's minimum.

Use the **left arrow key** to move the cursor until it is to the left of the function's minimum.

Press **ENTER** and this x value becomes the lower bound.

The calculator waits for you to select an **Upper Bound** for the interval containing the function's minimum.

Press the **right arrow key** at least twice to move the cursor until it is to the right of the function's minimum.

Press **ENTER** and this x value becomes the upper bound.

The calculator waits for you to select a **Guess** for the function's minimum.

Press the **left arrow key** to move the cursor close to the function's minimum.

Be careful not to move the cursor past the lower bound. The guess must be between the lower and upper bounds.

Press **ENTER** when the cursor is close to the minimum.

The calculator begins approximating the function's minimum. A moving vertical line segment is displayed in the upper right corner of the screen to indicate that the calculator is computing. When finished, the calculator displays the approximate minimum of the function within the interval defined by the lower and upper bounds. This solution is correct with an error of no more than 0.00001.

Example: Find the vertex of $y = x^2 - 3x - 2$. The vertex of this function is the lowest point on the graph. The graph appears in the standard viewing window.

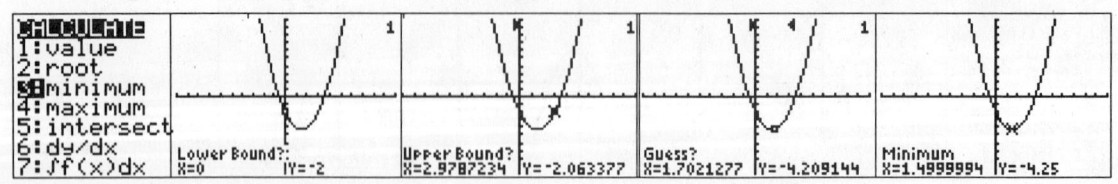

The vertex is actually (1.5, −4.25).

Working With Lists

Entering data into a list

A list of data can be entered, displayed, copied to another list, stored, sorted, used to graph families of curves, and in mathematical expressions. Data that has been entered in lists can be used to plot statistical data. The types of plots available include scatter plots, box-and-whisker plots, x-y plots and histograms. Calculations to fit the data entered in lists to one or more models is also possible. The resulting equations can then be stored in the Y = menu, graphed and traced.

From the Home Screen:

Press **STAT** **ENTER** to select **1:Edit** form the **STATEDIT** menu and to display the first three lists, L_1, L_2, and L_3.

With the lists displayed:

Position the cursor in the column with heading L_1 using the arrow keys. At the bottom of the screen you should see $L_1(1) =$. The 1 in the parentheses indicates that the cursor is at the first position in the list.

Enter each of the data elements.

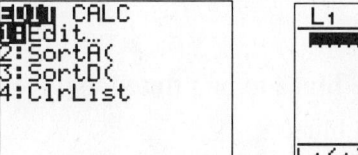

Press **ENTER** or **▽** after each element is entered.

Clearing data from a list

If data is already entered in a list you wish to use and you would like to erase the entire list:

Use the **△** , **▷** or **◁** keys and position the cursor in the column heading over the list name.

Press **CLEAR** **ENTER** to remove the existing data elements and clear the list.

The empty list is ready for entry of new data elements.

195

Changing numbers in an existing list

A number in a list can also be changed by positioning the cursor over that number and typing in the desired new number, followed by **ENTER** .

Inserting new numbers into an existing list

If you wish to insert a new number into an existing list without removing any of the numbers from the list first position the cursor where you wish to place the new number. Then put the calculator in **insert** mode by pressing **2nd** **DEL** . The abbreviation for **insert** appears above the **DEL** key in blue. A zero appears in the list where you wish to insert a new number. Type the new number and press the enter key.

Entering data into a list from the Home Screen

Data entered on the Home Screen using list notation and with each element followed by a comma can then be stored in one of the six lists.

From the Home Screen:

Press **2nd** **(** . This causes the brace to be printed. Notice that the left brace appears above the left parenthesis key in blue.

Enter the data elements separating each element with a comma.

Press **2nd** **)** after the last data element is entered. This prints the right brace which appears above the right parenthesis key in blue.

```
{                    {1,2,3,4,5}▮
```

196

Storing and viewing the newly-generated list:

After entering the list in braces on the home screen press `STO ▷` `2nd` `1`, then press `ENTER` to store the list in L_1. Notice that L_1 appears above the key with 1 on it, printed in blue.

Press `STAT` `ENTER` to select **1:Edit** from the **STATEDIT** menu and view the newly-generated list on the list screen.

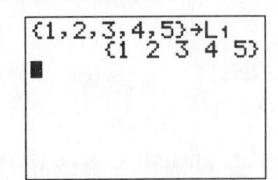

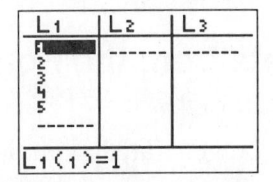

Pressing `STO` `2nd` `2`, and then `ENTER` would store the list in L_2.

Press `STAT` `ENTER` to select **1:Edit** from the **STATEDIT** menu and view the newly-generated list on the list screen.

Applying arithmetic operations to an existing list to create new list

A new list can be created either on the home screen and then stored in a blank list or on the list screen.

Generating new data on the list screen:

Press `STAT` `ENTER` to select **1:Edit** from the **STATEDIT** menu and to display L_1, L_2, and L_3.

Position the cursor on the column heading L_2 using the arrow keys.

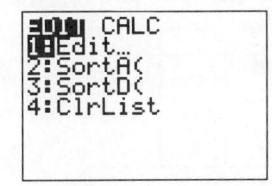

Create a new list L$_2$, by adding three to each element of L$_1$:

Type $L_1 + 3$ and press **ENTER**. The calculator adds 3 to each element of **L$_1$** and stores this new list in **L$_2$**.

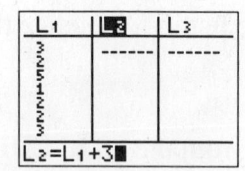

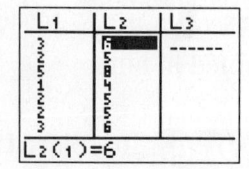

Generating new data from the Home Screen

A new list of data can be generated from the home screen using an existing list name. The data generated can be stored as a new list and viewed on the list screen.

From the Home Screen:

Press **STAT** **ENTER** to select the **EDIT** menu and to display L$_1$, L$_2$, and L$_3$.

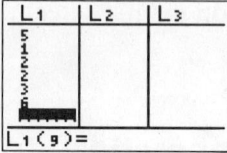

Create a new list L2, adding three to each element of an existing list L1:

Press **2nd** **QUIT** to return to the Home Screen.

Type $L_1 + 3$

Store and view the newly-generated list into L$_2$.

Press **STO** **2nd** **2**, then press **ENTER** to store the list in L$_2$.

Press **STAT** **ENTER** to select **1:Edit** from the **STATEDIT** menu and view the newly-generated list.

Use right and left arrows to view additional elements of the list on the Home Screen.

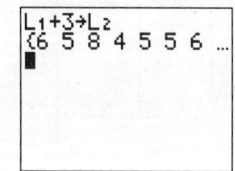

Generating a list of data using the sequence function in the LIST OPS menu

From the Home Screen access the **LIST OPS** menu:

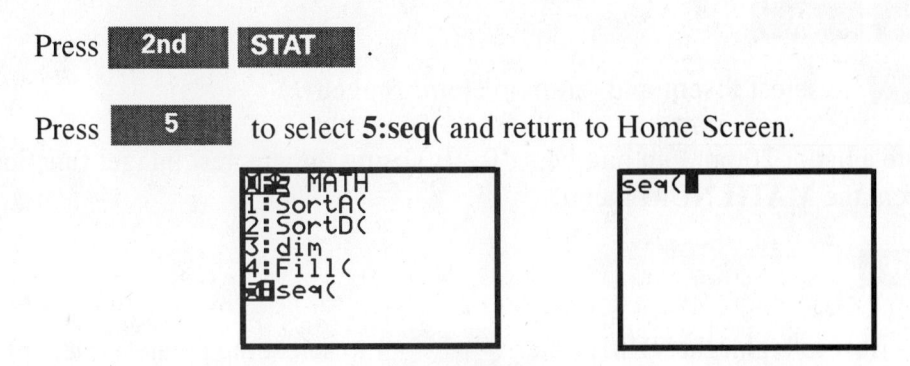

Press `2nd` `STAT` .

Press `5` to select **5:seq(** and return to Home Screen.

Type $x^2, x, 1, 10, 1$) to generate the squares of the numbers 1–10.

Press `ENTER` to display the list of elements on the Home Screen. Use left and right arrows to view additional list elements on the Home Screen.

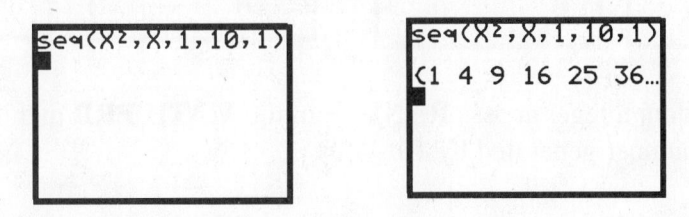

Note: The general format for entry of a sequence is:

  seq (generating rule, variable used in generating rule, initial value of the variable, ending value of the variable, increment of the variable)

Commas are necessary and must be included after each input for the sequence function.

To store and view the newly-generated list in L_1:

Press `STO` `2nd` `1` , then press `ENTER` to store the list in L_1.

Press `STAT` `ENTER` to return to the list screen.

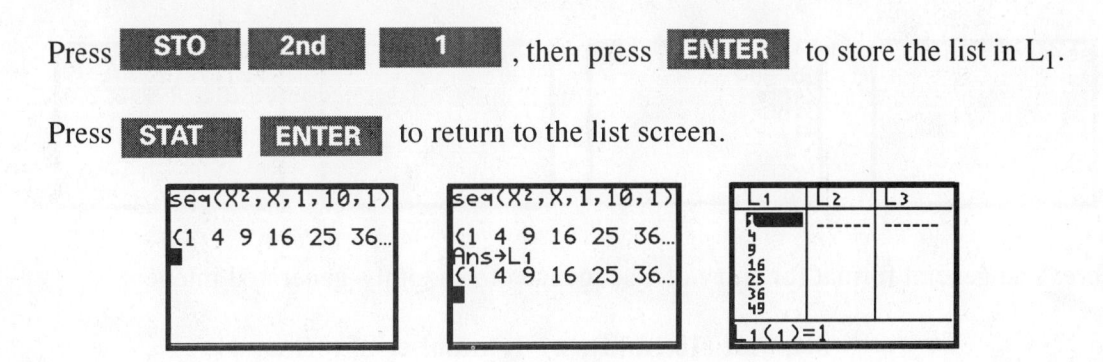

Generating a list of random numbers using sequences

From the Home Screen access the **LIST OPS** menu:

Press **2nd** **STAT**.

Press **5** to select **5:seq(** and return to Home Screen.

Generate and store a list of 20 random numbers (0 – 9) using the greatest integer function, **int**, accessed from the **MATH NUM** menu.

Press **MATH**.

Use the right arrow to highlight **NUM**. Enter **4** to select the greatest integer function *4: int*.

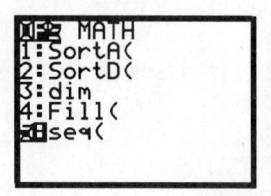

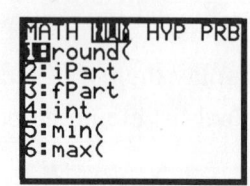

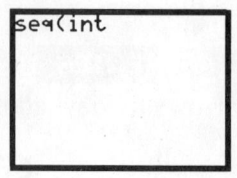

To generate a random integer access **RAND** from the **MATH PRB** menu and multiply the random decimal number generated by 10:

Enter **(10**

Press **MATH**.

Use the right arrow to highlight **PRB.**

Press **ENTER** to select the random seed generator.

Complete the sequence **), x, 1, 20, 1)**. The screen should look like the one below.

Press **ENTER** to display the list of elements on the Home Screen.

Use left and right arrows to view additional list elements on the Home Screen.

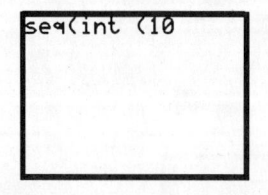

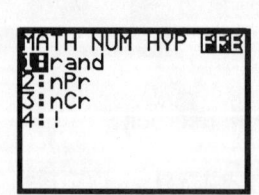

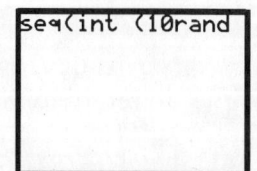

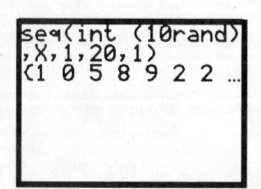

Note: The general format for entry of a sequence of randomly-generated integers is:

$$\text{seq (int (10 rand), } x, \text{ 1, number of terms, 1)}$$

Storing the newly-generated list in L_1:

Press **STO** **2nd** **1**, then press **ENTER** to store the list in L_1.

Press **STAT** **ENTER** to view the list on the list screen.

Use up and down arrows to view additional list elements on the list screen.

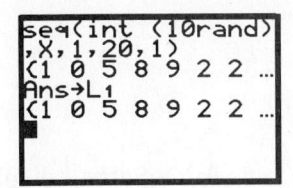

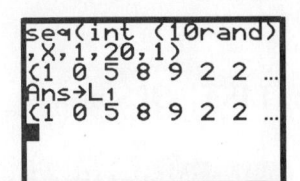

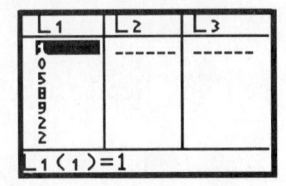

Adding a random number to each element of a list:

With the lists screen displayed:

Position the cursor on column heading L_2. At the bottom of the screen is displayed $L_2 =$

Complete the right-hand side by entering L_1 + int(10 RAND).

Press **ENTER** to display the list.

Use up and down arrows to view additional list elements.

NOTES:

- **RAND** generates and returns a random number between zero and 1 (0 < rand < 1).

- To generate more random number sequences, have individuals each store a different **integer** seed value in **RAND** first.

- If using a new set of TI-82s, have each person store an integer (social security numbers are a good choice) to generate random numbers.

- When you reset the TI-82 or store **0** to **RAND**, the factory-set seed value is restored.

Generating data using a sequence and an existing list (from the Home Screen)

A new list of data can be generated using an existing list with a sequence command from the Home Screen or with the lists displayed.

Using the list of twenty numbers randomly generated, return to the Home Screen.

Press **2nd** **QUIT**.

From the Home Screen access the **LIST OPS** menu:

Press **2nd** **STAT** **5** to select **5:seq(**. The Home Screen displays **seq(**.

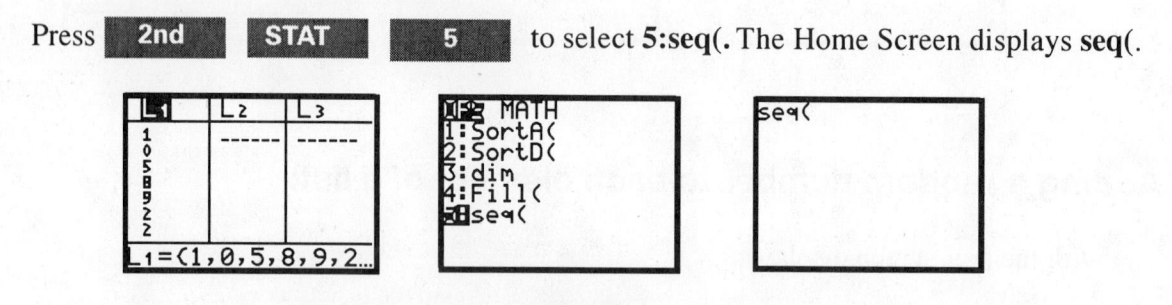

Type $L_1(x) + 3, x, 1, 20, 1)$ then press **ENTER** to display the list.

Press **STO** **2nd** **2**, then press **ENTER** to store the list in L_2.

Press **STAT** **ENTER** to return to the list screen.

Use up and down arrows to view additional elements of the list.

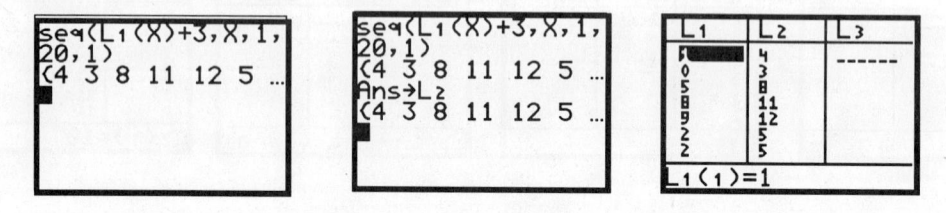

Note: The general format for entry of a sequence using an existing list is:

seq (L_{old} (x) with operation(s), x, 1, number of elements in the list, 1)

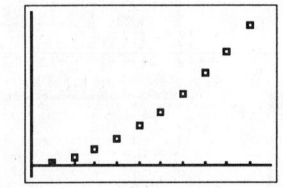

Constructing a scatter plot of data

A scatter plot is a two dimensional graph of ordered pairs and requires two lists of data values, one for the first coordinates and the second for the second coordinates. Scatter plots plot the data points from two lists, Xlist and Ylist, as ordered pairs. You can select one of three options for highlighting the points: as a box (□), a cross (+) or a dot (•). The number of data points in the Xlist and in the Ylist must be the same. The frequency of occurrence of each data value does not apply to scatter plots.

From the Home Screen:

Press **Y =** and turn off all selected Y= equations.

Press **STAT** **ENTER** to view the list screen and display L_1, L_2, and L_3.

If data has been previously entered in L_1, L_2, and L_3, clear existing data.

Enter the first set of data in L_1 and the second set of data in L_2.

Press **2nd** **Y =** to access the **STAT PLOTS** menu.

Select **1** for **Plot1**. Highlight **On** and press **ENTER**.

Use the **▽** and **▷** to highlight the first icon under the **Type:** options.

Press **ENTER**. This darkens the first icon which is the **scatter plot** icon.

Use the **▽** and **▷** keys to highlight L_1 in **Xlist**, L_2 in **Ylist**, and the square in the **Mark:** options.

Press **ENTER** after each selection.

Set a viewing window that contains all the data points and then display the scatter plot on the graphics screen.

Press **ZOOM** **9** to select **9:ZoomStat** from the **ZOOM** menu, set a viewing window and display the scatter plot.

203

Constructing a line plot of data

Enter the data in lists L₁ and L₂.

Press **2nd** **Y=** to access the **STAT PLOTS** menu.

Select **2** for **Plot2.** Highlight **On** and press **ENTER**.

Use the **▽** and **▷** to highlight the second icon in the **Type:** options.

Press **ENTER**. This selects the second type which is the **line plot** option.

Use the **▽** and **▷** keys to highlight **L₁** in **Xlist,** and **L₂** in **Ylist**.

Press **ENTER** after each selection

Select **ZoomStat** by pressing **ZOOM** **9**. This command sets a viewing window that contains all the data points and then displays the scatter plot.

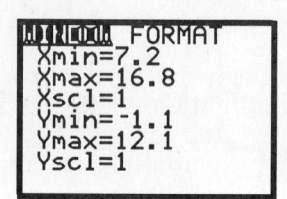

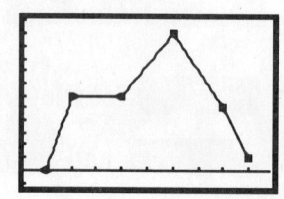

Turning off Stat Plots

When finished with them, it is a good idea to turn off the statistical plots. Not doing so often results in confusing graphs and error messages at a later time when you may be graphing functions entered in the Y= menu.

Press **2nd** **Y=** to access the **STAT PLOTS** menu.

Select **4** for **PlotsOff**. The calculator returns to the home screen.

Press **ENTER**. All statistical plots are now off.

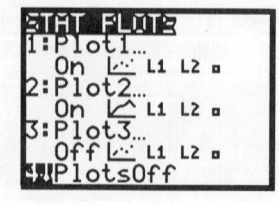

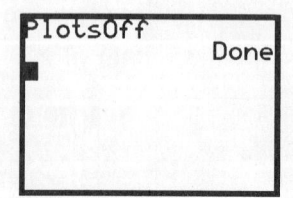

Finite differences and ratios using the sequence function

Finite differences can be calculated using the sequence function for data entered in a list. If the data consists of ordered pairs, enter the x-coordinates in the first list and the y-coordinates in the second list.

Example: Enter the following table into two list on the list screen and calculate the finite differences for the values in the second column:

X	S(x)
1	35,000.00
2	37,000.00
3	39,000.00
4	41,000.00
5	43,000.00
6	45,000.00
7	47,000.00
8	49,000.00
9	51,000.00
10	53,000.00

If the function is known and entered in the Y= menu, then the ordered pairs can be entered into two lists element by element or by using the function rule with the sequence command. In the table above, the left column is a list of sequential numbers (years) from 1 to 10; the right column is a list of the first ten years of salaries for a fictitious company named CD-R-Us, given an initial starting salary of $35,000, with annual pay raises of $2000 each subsequent year.

Enter the column of sequential years, X, as the first list, L_1, using the sequence function:

Press **STAT** **ENTER** to view the list screen and to display L_1, L_2, and L_3.

Position the cursor on the column heading L_1 using the arrow keys.

Press **2nd** **STAT** **5** to select **5:seq(**.

Type $x + 1, x, 0, 9, 1)$ to generate the list of numbers 1 – 10. Press **ENTER**.

205

Enter the column of salaries, S(X), as the second list, L_2:

Position the cursor on the column heading L_2 using the arrow keys.

Press `2nd` `STAT` `5` to select **5:seq(**

Type: $35000 + 2000(x-1), x, 1, 10, 1)$ to generate the list of salaries. Press `ENTER`.

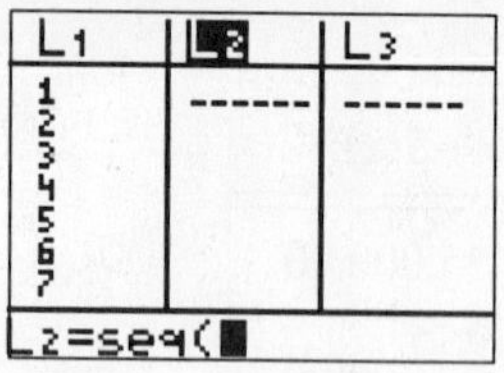

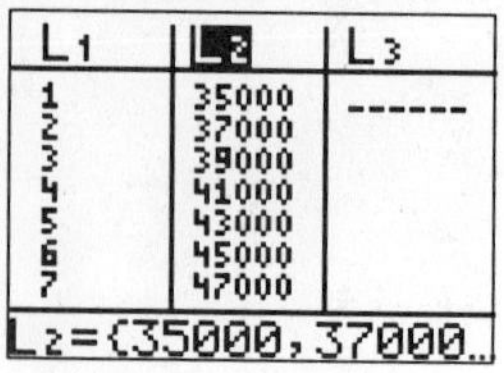

Generate the first finite differences of L_2 and store in L_3:

Position the cursor on the column heading L_3 using the arrow keys.

Press `2nd` `STAT` `5` to select **5:seq(**.

Type $L_2(x+1) - L_2(x), x, 1, 9, 1)$ to generate the list of first finite differences.

Press `ENTER`.

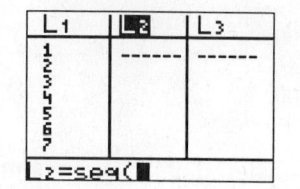

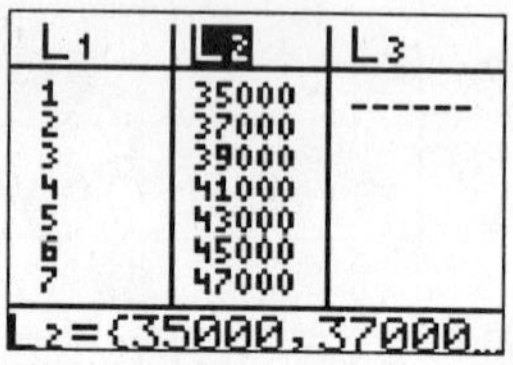

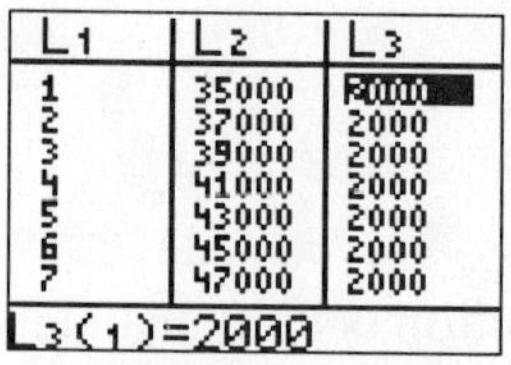

Generate the finite ratios of L_2 and store in L_4:

Position the cursor on the column heading L_4 using the arrow keys.

Press `2nd` `STAT` `5` to select **5:seq(**.

Type $L_2(x+1) / L_1(x), x, 1, 9, 1)$ to generate the list of finite ratios.

Press `ENTER`.

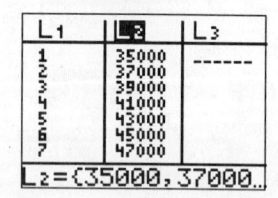

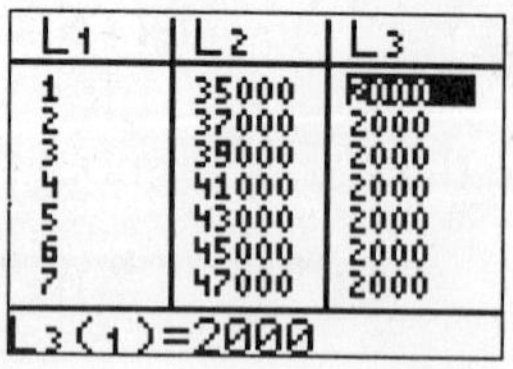

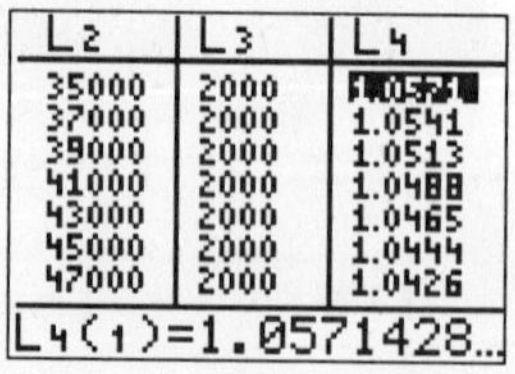

Note: The first finite difference is constant, an indication that the function is linear.

Using finite differences to select an appropriate regression equation

Data generated by an unknown sequence can be investigated using finite differences in order to select an appropriate regression equation and determine the function.

Example: Use finite differences to investigate the sequence

$$1, 3, 6, 10, 15, 21, 28, 36, 45, 55, \ldots$$

The first list consists of the positions of the known terms in the sequence; the second list consists of the terms that are given of the unknown sequence.

Position: 1, 2, 3, 4, 5, 6, 7, 8, 9, 10, ...
Seq term: 1, 3, 6, 10, 15, 21, 28, 36, 45, 55, ...

Enter the position numbers as the first list, using a sequence.

Press **STAT** **ENTER** to select the **EDIT** menu and to display L_1, L_2, and L_3.

Position the cursor on the column heading L_1 using the arrow keys.

Press **2nd** **STAT** **5** to select **5:seq(**.

Type $x + 1, x, 0, 9, 1)$ to generate the list of numbers 1 – 10.

Press **ENTER**.

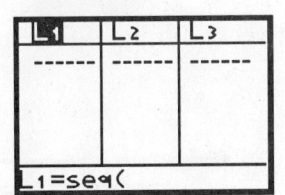

Enter the terms of the sequence as the second list, L_2, element by element.

Position the cursor in L_2 for the first entry.

Enter the terms of the sequence, element by element. Press **ENTER** after each term.

Generate the first finite different as L₃:

Position the cursor on the column heading **L₃** using the arrow keys.

Press [2nd] [STAT] [5] to select **5:seq(**.

Type $L_2(x+1) - L_2(x), x, 1, 9, 1)$ to generate the list of first finite differences.

Press [ENTER].

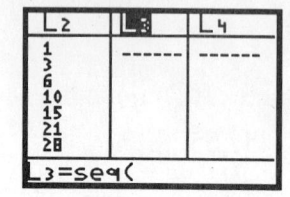

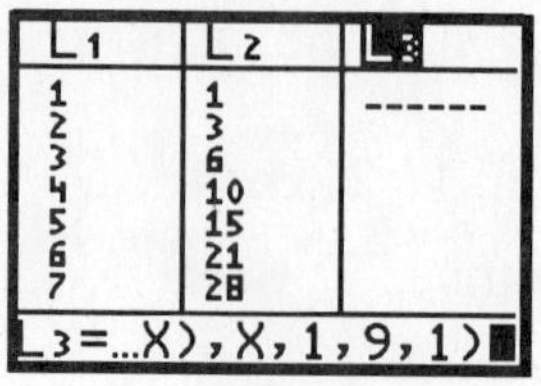

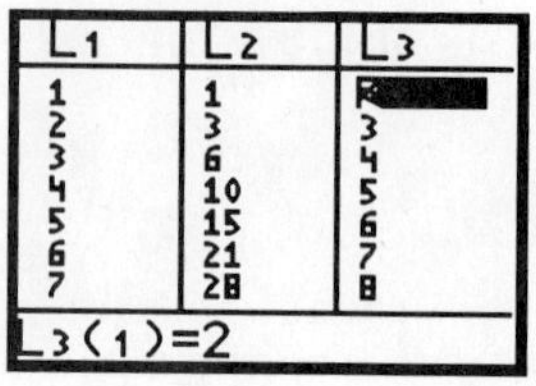

A linear model is indicated by the first finite difference being constant. If, however, the first finite difference is not constant, the function is not linear and you should continue to take finite differences.

Generate the second finite difference as L₄:

Position the cursor on the column heading **L₄** using the arrow keys.

Press [2nd] [STAT] [5] to select **5:seq(**.

Type $L_3(x+1) - L_3(x), x, 1, 8, 1)$ to generate the list of second finite differences.

Press [ENTER].

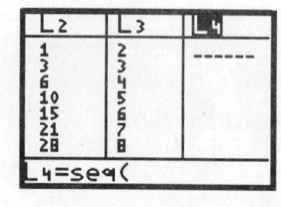

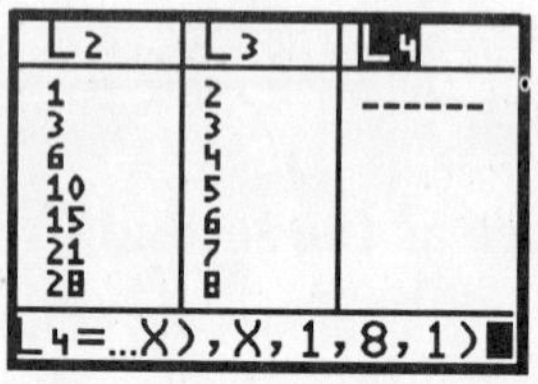

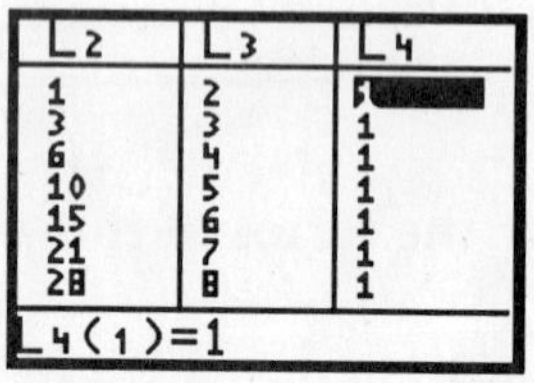

Note: The second finite difference is constant, an indication that the function is quadratic. The function can be determined using the quadratic regression model.

Creating a linear regression model

The TI–82 has the capability of creating the equation of a line that best fits a set of paired data. Such a line is called a linear regression model for the data.

Example: Enter the following paired data as lists L_1 and L_2.

x	1	2	7	9	12
y	4	11	23	32	45

Enter the paired data in two lists, L_1 and L_2.

Construct a scatter plot of the data (see index of procedures).

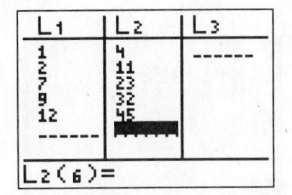

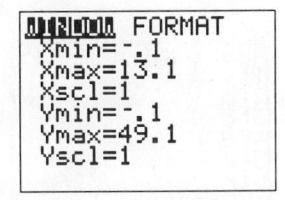

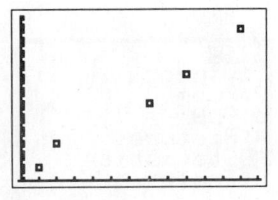

Press **STAT** **▷** to select the **CALC** menu.

Press **5** **ENTER** to select **LinReg**. The calculator returns to the home screen.

Press **2nd** **1** **,** **2nd** **2** **ENTER** to enter the names of the two lists, separated by a comma. The best–fit linear model is displayed

The value of a is the slope of the line.

The value of b is the vertical intercept of the line.

The value of r measures how well the line fits the data. The closer the absolute value of r is to one, the better the fit.

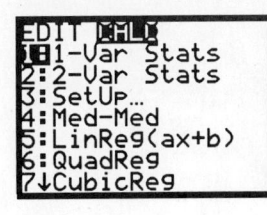

Store the regression model as one of the functions Y₁ to Y₀

Use "Creating a linear regression model for data" to create the best-fit linear equation.

Press [Y=] and place cursor on the function that will be assigned the regression model.

Press [VARS].

Press [1] to select **STATISTICS**.

Press [7] to select **RegEQ**.

The regression model appears in the Y= menu.

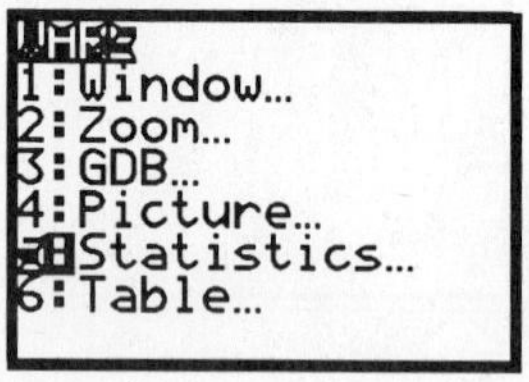

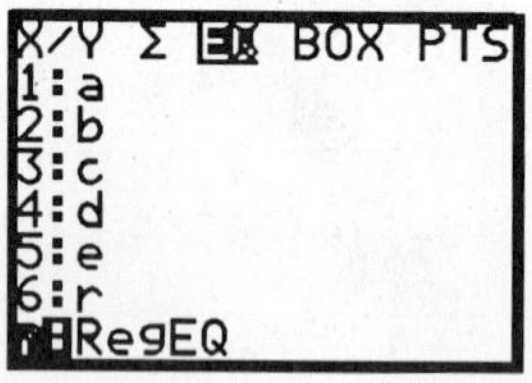

Graph the regression model

Store the regression model as a function using the **Y =** menu.

Adjust the viewing window if necessary.

Press [GRAPH] to display the regression model.

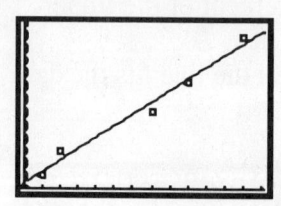

Creating a quadratic regression model

The TI–82 has the capability of creating the equation of a line that best fits a set of paired data. Such a line is called a regression model for the data.

The first step in creating a regression model is to construct a scatter plot of the data. A graph can be constructed using the data entered into two lists and the statistical plot capability of the calculator.

Construct a scatter plot of data

Enter the data in lists L_1 and L_2 as indicated below.

Press **2nd** **Y =** to access **STAT PLOT**.

Select **1** for **Plot1**.

Highlight **On** and press **ENTER**.

Use the arrow keys to move down to **Type** and highlight the first figure (scatter plot).

Press **ENTER** to turn on the **scatter plot.**

Repeat the process of using arrow keys to highlight L_1 under **Xlist**, and L_2 under **Ylist**.

Press **ENTER** after each selection is highlighted.

Set a viewing window that contains all the data points and then display the scatter plot.

Press **ZOOM** **9** to select **ZoomStat** and set a viewing window.

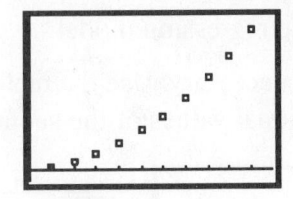

Select an appropriate regression model.

The data points displayed on the scatter plot give the appearance of being quadratic. A good first choice would be the quadratic regression model.

Press **STAT** **▷** to select the **CALC** menu.

Press **6** **ENTER** to select **QuadReg**. The Home Screen is displayed.

Press **2nd** **1** **,** **2nd** **2** **ENTER** to enter the names of the two lists, separated by a comma. The best–fit quadratic model is displayed.

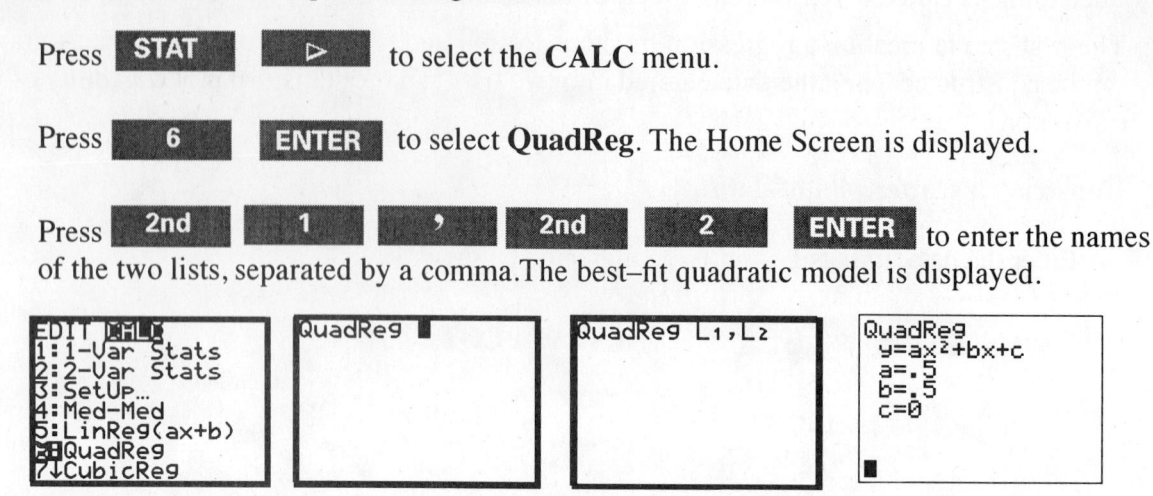

The value of a is the coefficient of the quadratic term. The value of b is the coefficient of the linear term and the value of c is the vertical intercept and constant term.

Store the regression model as one of the functions Y_1 to Y_0.

Press **Y=** and place cursor on the function assigned the regression model.

Press **VARS** **5** to select **STATISTICS**.

Press **7** **ENTER** to select **RegEq**.

The regression model appears in the **Y=** menu.

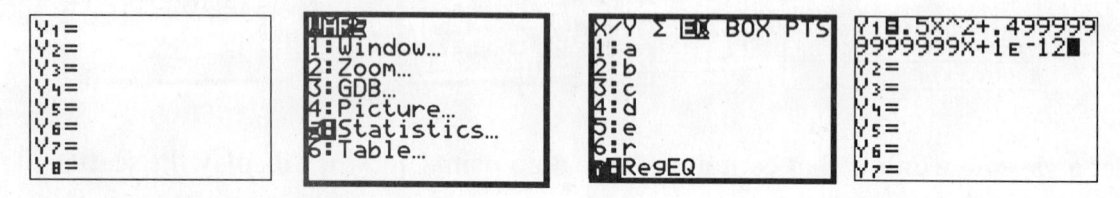

Graph the regression model.

Press **GRAPH** to display the regression model.

Adjust the viewing window if necessary. Use the table generated by the equation entered in the Y = menu to view additional values of the sequence.

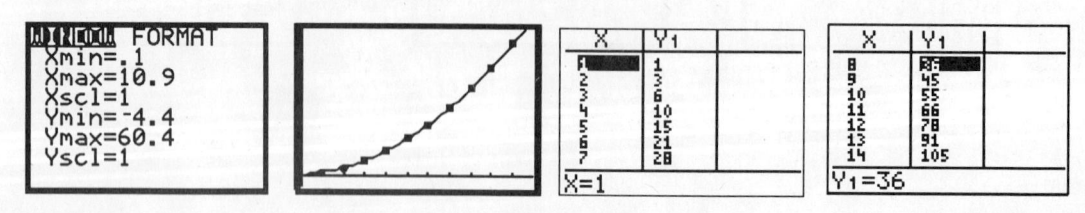

Transferring data between calculators: using LINK

Connect the two calculators using the connection cable.

Person receiving the data:

Press [2nd] [X, T, θ] to select **LINK**.

Press [STAT] [▷] to select **RECEIVE** and press [ENTER]. Now just wait.

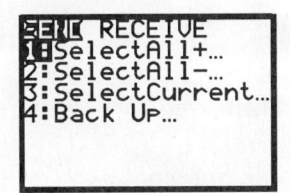

```
SEND RECEIVE
1:SelectAll+...
2:SelectAll-...
3:SelectCurrent...
4:Back Up...
```

```
SEND RECEIVE
1:Receive
```

Person sending the data:

Press [2nd] [X, T, θ] to select **LINK**.

Press [2] to choose **SelectAll–**.

Press [▽] [ENTER] to select each item you wish to send. A small black box appears next to each item that you select.

Press [▷] to select **TRANSMIT** and press [ENTER].

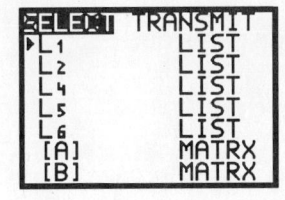

```
SELECT TRANSMIT
▶L1      LIST
 L2      LIST
 L4      LIST
 L5      LIST
 L6      LIST
 [A]     MATRX
 [B]     MATRX
```

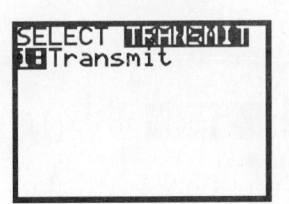

```
SELECT TRANSMIT
1:Transmit
```

Disconnect calculators.

Entering and displaying a matrix

A matrix is a two-dimensional array of rows and columns. The dimensions of a matrix, the number of rows and the number of columns, is designated with the row number preceding the column number. The TI-82 stores as many as five matrices using the matrix variables [A], [B], [C], [D], and [E]. You can display, enter, or edit a matrix in the matrix editor.

Example: Enter the 3 x 3 matrix $\begin{bmatrix} 2 & -5 & 3 \\ 0 & 1 & -2 \\ -4 & 7 & 6 \end{bmatrix}$.

Set the dimension of the matrix. If you are using matrices to solve a system of equations, the number of equations is the row number and the number of variables is the column number.

Press **MATRIX** and arrow to **EDIT**. Type **1** to select matrix [A].

Enter the size of the matrix (in this case, 3 x 3) followed by the matrix elements.

Press **2nd** **MODE** to return to the Home Screen.

Press **MATRIX** **1** **ENTER** to display matrix [A] in the Home Screen.

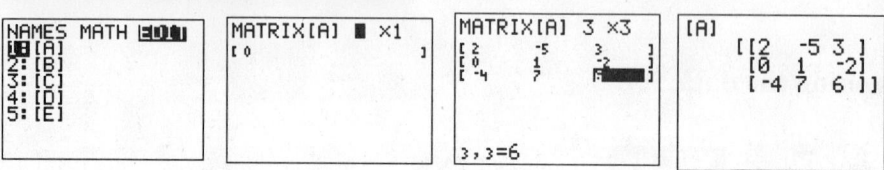

Calculating the determinant of a matrix

Example: Find the determinant of the 3 x 3 matrix $\begin{bmatrix} 2 & -5 & 3 \\ 0 & 1 & -2 \\ -4 & 7 & 6 \end{bmatrix}$.

Enter the matrix, if you haven't done so, as matrix [A]. Return to the Home Screen.

Press **MATRIX** and arrow to **MATH**.

Type **1** to select **det**, which represents a determinant.

The calculator returns to the Home Screen and displays **det**.

Press **MATRIX** **1** to display [A] after **det**.

Press **ENTER** to see the determinant of matrix [A].

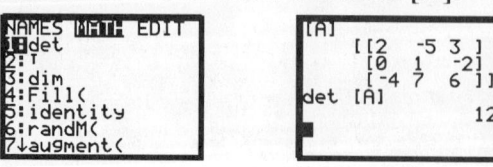

Deleting a matrix from memory

Press **2nd** **+** to access the **MEM** menu.

Press **2** to display the 2: Delete options.

Press **4** to display the matrix deletion options. Select the matrix you wish to delete.

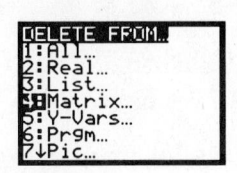

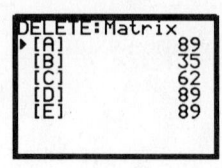

```
MEMORY
1:Check RAM...
2:Delete...
3:Reset...
```

```
DELETE FROM..
1:All...
2:Real...
3:List...
4:Matrix...
5:Y-Vars...
6:Prgm...
7↓Pic...
```

```
DELETE:Matrix
▶[A]      89
 [B]      35
 [C]      62
 [D]      89
 [E]      89
```

Press **ENTER** to delete the selected matrix.

Finding the inverse of a square matrix

Example: Find the inverse of the 3 x 3 matrix $\begin{bmatrix} 2 & -5 & 3 \\ 0 & 1 & -2 \\ -4 & 7 & 6 \end{bmatrix}$.

Enter the matrix, if you haven't done so, as matrix [A].

Return to the Home Screen.

Press **MATRIX** **1** **x⁻¹** **ENTER** to display the inverse of matrix [A] in decimal form.

The left and right arrow keys can be used to see all elements of the 3 x 3 inverse matrix.

```
[A]⁻¹
[[1.666666667 4...
 [.6666666667 2...
 [.3333333333  ...
```

```
[A]⁻¹
...4.25  .58333333...
...2     .33333333...
...5     .16666666...
```

Press **MATH** **1** **ENTER** to display the inverse matrix in fractional form.

```
...4.25  .58333333...
...2     .33333333...
...5     .16666666...
Ans▶Frac
[[5/3  17/4  7/12...
 [2/3  2     1/3  ...
 [1/3  1/2   1/6  ...
```

215

Solving linear systems using matrix inverses

A system of linear equations can be solved by entering the coefficients as elements in an augmented matrix, then finding the solution using matrix row operations. If the coefficient matrix is square (same number of rows and columns) and the determinant is not equal to zero, you may solve the system by finding the product of the inverse of the coefficient matrix and the constant matrix.

Example: Solve the system
$$\begin{aligned} 2x - 5y + 3z &= 7 \\ y - 2z &= -3 \\ -4x + 7y + 6z &= 1 \end{aligned}.$$

The coefficient matrix is $[A] = \begin{bmatrix} 2 & -5 & 3 \\ 0 & 1 & -2 \\ -4 & 7 & 6 \end{bmatrix}$ and the constant matrix is $[C] = \begin{bmatrix} 7 \\ -3 \\ 1 \end{bmatrix}$.

Enter the coefficient matrix as matrix[A], if you haven't done so.

Press **MATRIX** and arrow to **EDIT**.

Type **3** to select matrix [C].

Enter the size of the matrix (in this case, 3 x 1) followed by the matrix elements.

Press **2nd** **MODE** to return to the Home Screen.

The determinant of [A] is 12. The system has one solution since the determinant of the coefficient matrix is not zero.

If the linear system has a unique solution, the computation $[A]^{-1}[C]$ will return a column matrix that contains the solution to the system.

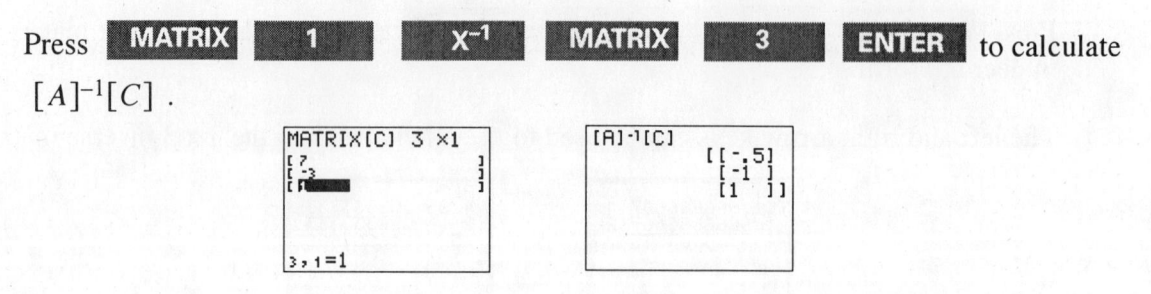

Press **MATRIX** **1** **X⁻¹** **MATRIX** **3** **ENTER** to calculate $[A]^{-1}[C]$.

The 3 x 1 solution matrix displays the solution to the system
$$\begin{aligned} 2x - 5y + 3z &= 7 \\ y - 2z &= -3 \\ -4x + 7y + 6z &= 1 \end{aligned}.$$

Reading down the matrix, the solution is $x = -0.5$, $y = -1$, and $z = 1$.

Index of TI-82 Procedures

A

Access functions using the 2nd key 228
Add a random number to each element of a list 259
Adjust the display contrast 223
Adjust the viewing window 241
Apply arithmetic operations to an existing list to create new list 255
Arithmetic computations 229
ASK mode 238

C

CALC 248
Calculate the determinant of a matrix 272
Calculations 229
Change the mode 224
Change the view window 241
Clear data from a list 253
Connected Mode for graphs 245
Construct a scatter plot of data 261
Construct a line plot of data 262
Control the TRACE option 243
Create a linear regression model 267
Create a quadratic regression model 268

D

Darken or lighten screen 223
Delete a matrix from memory 272
Determinant of a matrix 272
Discrete graph display 245
Display coordinates of points on a graph 242
Display a discrete graph 245
Display a table of values 234
Dot Mode for graphs 245

E

Edit an existing list 254
Edit a previously-entered expression 230
Enter and display a matrix 271
Enter data into a list 253
Enter data into a list from the Home Screen 254
Enter expressions in the Y = menu 233
Evaluate a function 236, 237
Exponentiation 229

F

Factorial function 233
Find zeros of a function 248
Find intersection points 249
Find a function's maximum 250
Find a function's minimum 251
Finite differences 263, 265
Finite ratios 263
Friendly view windows 243
Function evaluation: storing input 236
Function evaluation using Ask mode 238
Function evaluation using function notation 237

G

Generate data using an existing list 260
Generate data from the Home Screen 256
Generate new data on the list screen 255

Generate a list of data using the sequence function 257
Generate a list of random numbers 258
Graph a function 239
Graph the regression model 268
Graph the solution interval to inequalities in one variable 247
Graphics screen 225

H
Home screen 225

I
Important keys and their function 227
Inequalities: graphing a solution interval 247
Insert new numbers into an existing list 254
INT function 231
Intersection points: finding 249
Inverse of a Square Matrix 273

K
Keyboard organization 222

L
Linear regression models 267
LIST OPS menu 257
List screen 226
Lists of data 253

M
Maximum of a function 250
Matrix: determinant of 272

Matrix: inverses of a square 273, 274
Menus 224
Menu screens 224
Minimum of a function 251
Mode menu 245

N
Negative numbers 229

O
Obtain fractional results 231

P
Parametric graphing 246

Q
Quadratic Regression models 268

R
Raise a number to a power 229
Random number generation 259
RAND function 232
Reciprocal function 230
Regression equation 268
Regression models: graphing 268
Regression models: linear 267
Regression models: quadratic 268

S
Scientific notation (3.46E11 means) 230
Select a menu item 224
Setting a view window 243
Standard view window 240

STAT PLOT 262
Store a function input 236
Store the regression model 210
Store a value for a variable 235
Store a newly generated list 257

T

Table screen 226
TRACE 242
Transfer data between calculators 270
Turn on/off Stat Plots 262
Turn the calculator on/off 222

V

View a newly generated list 257
View window adjustments 241
View window: standard 240
View window: friendly 243

Y

Y = menu 225, 233

W

Working With Lists 253

Z

Zeros of a function 248
ZOOM options to adjust the view window 243

TI-83
Graphing Calculator
Reference Manual

TI-83 REFERENCE MANUAL

The TI-83 represents an upgrade to the TI-82 graphics calculator. Many of the procedures, screens, and menus on the two calculators are similar. However, some procedures are located under different menus or have different keystrokes. This part of the manual will describe the TI-83 operations.

Turn the calculator on and off

Press **ON** to turn the calculator on. If nothing appears on the screen, adjust the display contrast (see below). If you still see nothing, check the batteries.

Press **2nd** **ON** to turn the calculator off. The TI-83 is equipped with an automatic power down feature which turns the calculator off if no keys have been pressed for about five minutes.

Keyboard organization

The keys are grouped by both color and location for ease of use.

The keys are color–coordinated.

To choose a command that is in yellow and appearing above a key, press the yellow **2nd** key first. By convention, any key that is in yellow will appear with black, rather than white, letters in this manual. Notice that the cursor changes from a blinking solid black rectangle to a rectangle with a light upward pointing arrow inside when the second key is depressed.

To choose a letter or command which appears in green and above a key, press the green **ALPHA** key first. Notice that the cursor changes from a blinking solid black rectangle to a rectangle with a light A inside when the Alpha key is depressed.

Keys by row beginning at the top of the calculator:

Row 1: Graphing and table keys.

Y= **WINDOW** **ZOOM** **TRACE** **GRAPH**

Rows 2 and 3: Editing keys.

Rows 3 and 4: Advanced functions accessible through pull-down menus.

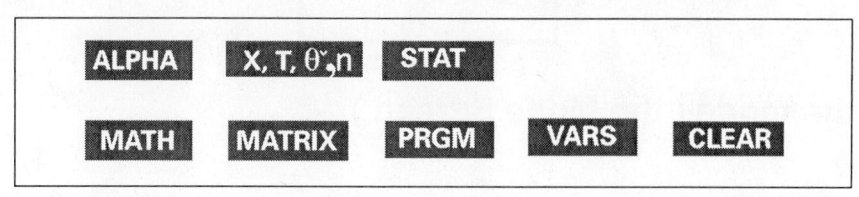

Rows 5–10: Scientific calculator keys.

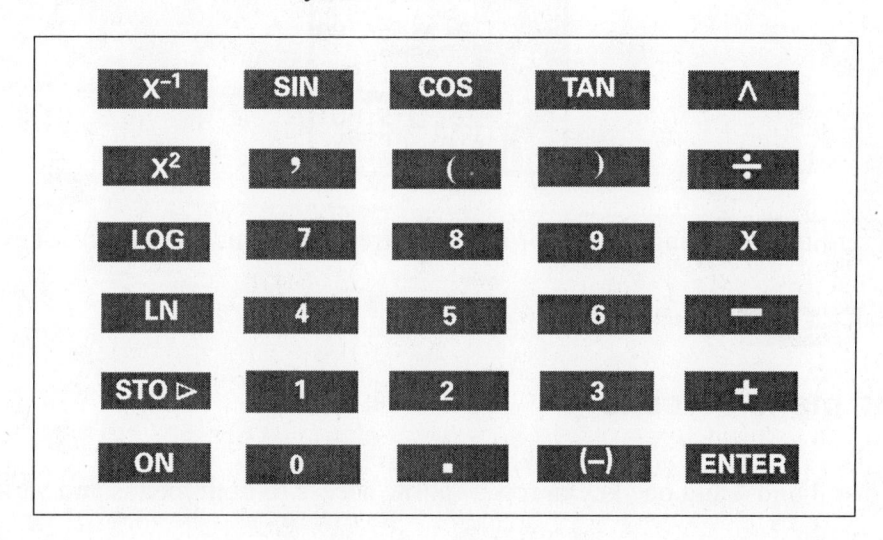

Adjust the display contrast (darken or lighten screen)

Press and release the [2nd] key.

Press and hold [▲] to increase the contrast (darken the screen).

Press and hold [▼] to decrease the contrast (lighten the screen). A black rectangle with a number inside appears in the upper right hand corner of the screen when darkening or lightening the screen. The number indicates the relative darkness of the screen. 9 is corresponds to the darkest setting while 0 indicates that the screen is as light as it will get. If you find it necessary to have the darkness set at 8 or 9 in order to see the screen then it is time to replace the four triple A batteries. In addition, the TI-83 displays a message that the batteries need changing when they are weak.

The calculator can be returned to factory settings by pressing [MEM]. Since MEM is in yellow above the "+" key, access MEM by pressing [2nd] [+]. Select item number **5:Reset** to reset the calculator to factory settings.

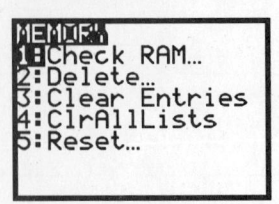

```
MEMORY
1:Check RAM...
2:Delete...
3:Clear Entries
4:ClrAllLists
5:Reset...
```

Changing the mode (the Mode Screen)

Press [MODE] to customize the calculator settings. Activated settings are highlighted.

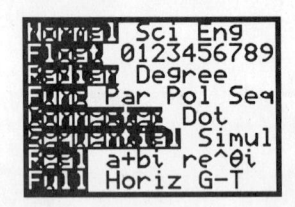

```
Normal Sci Eng
Float 0123456789
Radian Degree
Func Par Pol Seq
Connected Dot
Sequential Simul
Real a+bi re^θi
Full Horiz G-T
```

Change a setting by moving the cursor to the desired setting using the arrow keys.

Press [ENTER] to select a setting.

Menus and menu screens

Keys in rows 3 and 4 and one key in row 5 allow access to both menus and variable values.

Selecting a menu item

Press the key(s) corresponding to the menu to be displayed. For example, pressing the [MATH] opens the first menu shown. The other menus shown are accessed by using the right or left cursor controls to highlight the menu name that appears at the top of the screen.

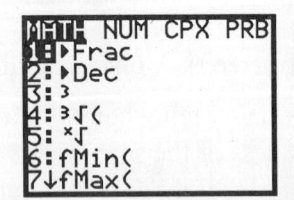

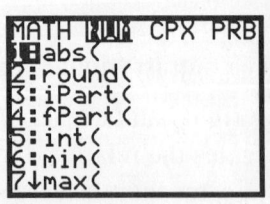

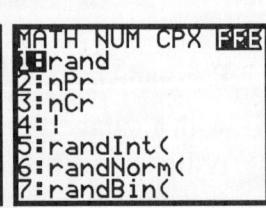

```
MATH NUM CPX PRB
1:▶Frac
2:▶Dec
3:³
4:³√(
5:ˣ√
6:fMin(
7↓fMax(
```
```
MATH NUM CPX PRB
1:abs(
2:round(
3:iPart(
4:fPart(
5:int(
6:min(
7↓max(
```
```
MATH NUM CPX PRB
1:conj(
2:real(
3:imag(
4:angle(
5:abs(
6:▶Rect
7:▶Polar
```
```
MATH NUM CPX PRB
1:rand
2:nPr
3:nCr
4:!
5:randInt(
6:randNorm(
7:randBin(
```

Select a menu item either by using the arrow keys to highlight the desired choice and then pressing [ENTER] ; or if a menu item is numbered, press the number of the menu item.

Home screen

The home screen is used for calculations.

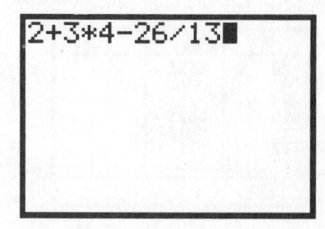

```
2+3*4-26/13█
```

You may return to the home screen at any time by executing the QUIT command. This is done by pressing **2nd** **MODE**. Notice that QUIT appears in yellow above the MODE key.

Graphics screen

The graphics screen is used to display graphs of functions that are entered into the Y= screen or of data which are entered in the list screen and plotted using the **STAT PLOTS** menu.

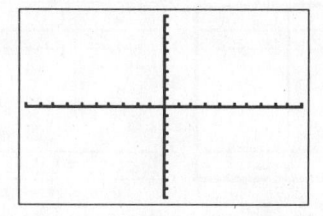

The part of the coordinate plane that is displayed is set using the window screen to be discussed later in the manual. To show the graphics screen press **GRAPH**.

Y= menu screen

The Y= menu screen is used to enter algebraic representations of functions. The graphs of these functions may be displayed on the graphics screen and input output tables generated from these functions may be displayed on the table screen. Ten different functions may be entered in this menu.

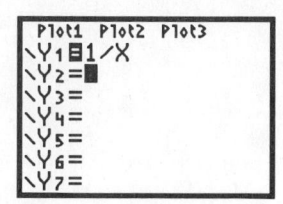

```
Plot1 Plot2 Plot3
\Y1=1/X
\Y2=█
\Y3=
\Y4=
\Y5=
\Y6=
\Y7=
```

Press **Y=** to display this screen.

225

Table screen

This screen displays a table of inputs and outputs for each active user-defined function.

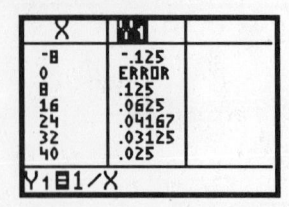

Choose **TABLE** to display this screen by pressing **2nd** **GRAPH**. Notice the word TABLE appears above the GRAPH key printed in yellow.

List screen

This screen displays lists where users may enter data. There are six lists in memory but it is possible to create additional list and name the new list with words that are up to five letters long.

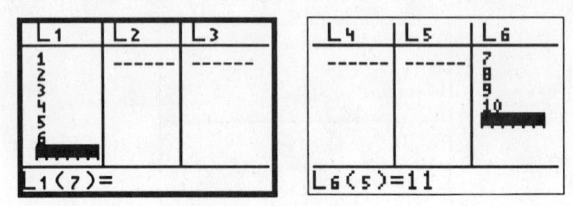

Press **STAT** and then press **1** to choose the Edit item on this menu. This is the **STAT EDIT** menu.

This shows the list screen and allows you to edit the lists. Press the right cursor control to show more list.

Important keys and their function

Keystroke	Function Description
ON	Turns on calculator; interrupts graph being sketched or other calculations.
CLEAR	Clears the text screen; deletes functions when in the Y= screen.
ENTER	Executes a command; moves cursor to next line.
(−)	The unary operation of opposing (not used for subtraction).
MODE	Displays mode screen to view current operating mode.
DEL	Deletes character at the cursor.
^	Symbol used for exponentiation
Y =	Accesses screen where up to ten functions may be entered and stored.
X, T, θ, n	Enters the variable x in function mode; enters the variable t in parametric mode; enters Θ in polar mode; enters n in sequence mode.
MATH	Accesses math, number, complex and probability menus and submenus.
2nd	Accesses functions and menus printed in yellow.
ALPHA	Accesses letters and symbols printed in green to the right and above many keys.
STO ▷	Used to assign a numeric value to a variable or store a list.
GRAPH	Displays the graphics screen.
WINDOW	Accesses menus for setting the viewing window.
ZOOM	Accesses ZOOM menu where various adjustments can be made to the viewing window.
TRACE	Activates trace mode and displays graphics screen and cursor. Use the left or right arrow keys to trace along a graph. Use the up and down arrow keys to move the trace cursor from one graph to another.

Accessing functions using the [2nd] key
(commands written in yellow, above and left of center of the keys).

Editing Functions Accessed with the [2nd] key.

Function	Keystrokes	Description
OFF	2nd ON	Turns the calculator off.
ENTRY	2nd ENTER	Allows the last line entered to be edited and re-executed.
INS	2nd DEL	Allows insertion of character(s) to the left of the cursor.
LINK	2nd STAT	Allows transfer of data or programs between two TI-83 or 82 calculators.
QUIT	2nd MODE	Exits current screen; returns calculator to home screen.

Table Functions Accessed with the [2nd] key.

Function	Keystrokes	Description
TABLE	2nd GRAPH	Displays a table of x–values and corresponding output values for functions stored in the **Y=** menu.
TblSet	2nd WINDOW	Accesses the TABLE SETUP menu where the minimum table value and the table increment are set.
STATPLOT	2nd Y=	Accesses the STAT PLOT menu where statistical data may be graphically displayed using a scatter plot, line plot, box–and–whisker plot or histogram.

Calculations

Calculations are done in the home screen. If you are not in the home screen, choose **QUIT** which is accessed by pressing **2nd** **MODE**.

Enter expressions as you would write them. Observe the algebraic order of operations.

After you complete typing an expression, press **ENTER** to obtain the answer.

If you make a mistake during entry, use the arrow keys, **INS** (for insert), or **DEL** (for delete) to edit your expressions.

Raising a number to a power **^**

Enter an exponent first entering the base and then pressing the **^** key followed by the exponent. For example, 4^5 should appear on the home screen as shown below. Press **ENTER** to display the value of the expression.

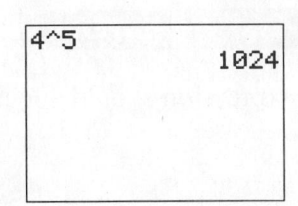

```
4^5
         1024
```

Negative numbers **(−)**

Enter a negative number using the *gray* key **(−)** (located to the left of the **ENTER** key), not the blue key, which is reserved for the operation of subtraction.

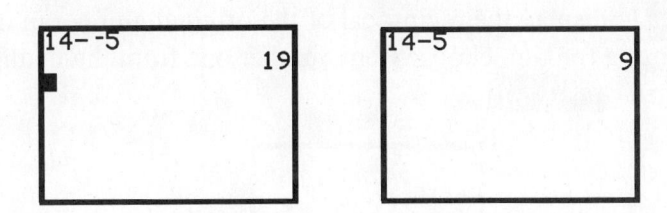

```
14−-5
           19
```
```
14−5
            9
```

Scientific notation (3.46E11 means 3.46×10^{11})

This notation uses powers of ten to represent very large and very small numbers. Numbers are written as a decimal times a power of 10.

To enter on the calculator, replace the **X 10** with **EE**. The exponent appears immediately to the right of the E in the display. The display, 3.46E11, actually represents 3.46×10^{11}.

Enter **EE** by pressing [2nd] [,]. Only one E appears on the screen to indicate that the next number entered will be accepted as a power of ten.

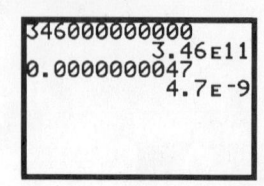

```
346000000000
           3.46E11
0.0000000047
           4.7E-9
```

Editing a previously-entered expression [2nd] [ENTER]

The last expression entered in the calculator can be displayed on the screen and edited. Execute the **Entry** command to display the last expression entered by pressing the following keys.

Choose [ENTRY] by pressing [2nd] [ENTER].

A copy of the previously entered expression should appear on the screen.

This expression may be edited.

The new expression may be executed by pressing [ENTER].

Reciprocal function [x⁻¹]

To find the reciprocal of a number first enter the number whose reciprocal you wish to find on the home screen. If the number is a fraction it must be enclosed in parentheses.

Press [x⁻¹].

Press [ENTER] to display the reciprocal of the original number in decimal form. To display the answer as a fraction choose item number one from the math menu by pressing [MATH] then [1].

```
(7/3)⁻¹
          .4285714286
Ans▶Frac
                 3/7
```

INT function

The greatest integer function (**int**) returns the largest integer less than or equal to a number, expression, list or matrix. For nonnegative numbers and negative integers, the **int** value is the same as **iPart**. For negative noninteger numbers **int** returns a value one integer less than **iPart**.

Press **MATH**.

Use the **▷** key to highlight the **NUM** menu.

Enter **5** or use the **▽** to highlight **5** and press **ENTER** to select the greatest integer function **5: int(**.

The calculator returns to previous screen and inserts "**int (**" at cursor. Follow the function name with an input value and close the parentheses. The function returns an output which is the largest integer that does not exceed the input.

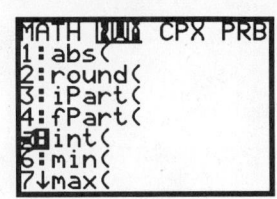

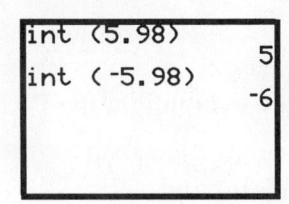

```
MATH NUM CPX PRB
1:abs(
2:round(
3:iPart(
4:fPart(
5:int(
6:min(
7↓max(
```

```
int (5.98)
            5
int (-5.98)
           -6
```

Obtaining fractional results

If a computation results in a decimal output, the result can often be converted to a fraction by using the **Frac** command.

Press **MATH** following a computation to open the **MATH MATH** menu.

Press **1** to select **Frac**. The calculator returns to previous screen.

Press **ENTER** and the calculator outputs a fraction.

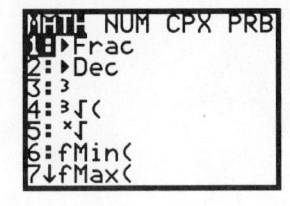

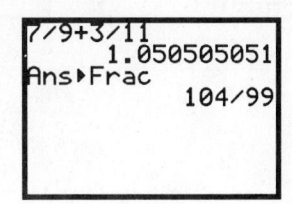

```
MATH NUM CPX PRB
1:▶Frac
2:▶Dec
3:³
4:³√(
5:ˣ√
6:fMin(
7↓fMax(
```

```
7/9+3/11
          1.050505051
Ans▶Frac
             104/99
```

231

RAND function

The calculator contains a random number generator that will return random numbers greater than zero and less than one.

The random number generator should be seeded initially so that each calculator will generate a different sequence of numbers. To control a random number sequence, first store an integer seed value in **rand**. The factory-set seed value is used whenever you reset the TI-83 or when you store **0** to **rand**.

Start in the home screen.

Type a favorite whole number. This number acts as the seed.

Press **STO▷**. This is the key used for storing a value in the calculator.

Press **MATH**.

Use the **◁** key to highlight the **PRB** menu.

Press **1** to select **rand**.
The calculator returns to the home screen and inserts the **rand** function.

Press **ENTER** to complete the seeding process.

Recall **rand** by pressing **MATH**, arrowing to **PRB**, and selecting **1**.

Press **ENTER** to see the first random number.

Continually pressing **ENTER** will generate additional random numbers.

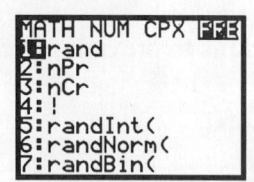

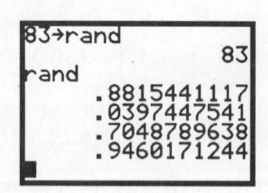

```
MATH NUM CPX PRB
1:rand
2:nPr
3:nCr
4:!
5:randInt(
6:randNorm(
7:randBin(
```

```
83→rand
                83
rand
        .8815441117
        .0397447541
        .7048789638
        .9460171244
```

Factorial function

The factorial of a whole number is the product of all whole numbers less than or equal to the given whole number.

Type a whole number.

Press **MATH**.

Use the **◁** key to highlight the **PRB** menu.

Press **4** to select **!** which represents factorial.

The calculator returns to the home screen and inserts the **!** after the number you typed.

Press **ENTER** to see the result of the factorial.

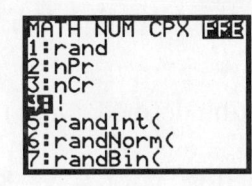

Entering expressions in the Y = menu **Y=**

When you enter a function the = symbol is highlighted to show the function is selected to be graphed or to display outputs on the table screen. The highlighting may be toggled on and off by using the arrow keys to move the cursor to the = symbol and then pressing **ENTER**.

Define the function:

Press **Y=** to access the function entry screen.

Type the expression for the function after $Y_1 =$. This defines the function as Y_1.

Select the *int* function:

Example: Enter the function $f(x) = x - 8int(x/8)$ as Y_1:

Press **Y=** to access the function entry screen.

Enter the function expression **x − 8** at the cursor.

Press **MATH**.

Use the **▷** key to highlight the **NUM** menu.

233

Press the **4** key or use the [▽] to highlight **4** and press [ENTER] to select the greatest integer function **4: int**.

Complete the entry of the function by entering (*x* / 8) then press [ENTER].

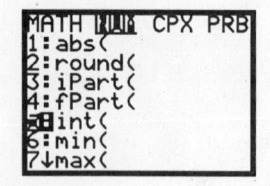

Functions may be removed from this menu by pressing [CLEAR].

Up to ten functions can be defined and accessed simultaneously.

Displaying a table of values from a function in the Y= menu

The TI–83 will display a table of values for a function defined under the Y = menu.

Example: Display a table of values for the function $y(x) = x - 8int\left(\frac{x}{8}\right)$

First Define the function:

Press [Y=] to access the function entry screen and enter *x - 8 int (x/8)* as Y_1.

Press [2nd] [WINDOW] to access [TblSet], the table set up menu.

Set the beginning table value and the increment:

Enter a beginning table value for **TblStart** at the flashing cursor.

Press the [▽] to place the cursor on the number next to Δ**Tbl** = to change the increment value.

ΔTbl represents the increment on the input values. The default value is **1**. If you wish a different increment, enter that value.

Check to see that **Auto** is highlighted for both Indpnt and Depend. If not, use arrow keys to place cursor on **Auto** and press [ENTER] to select it.

234

To view the table:

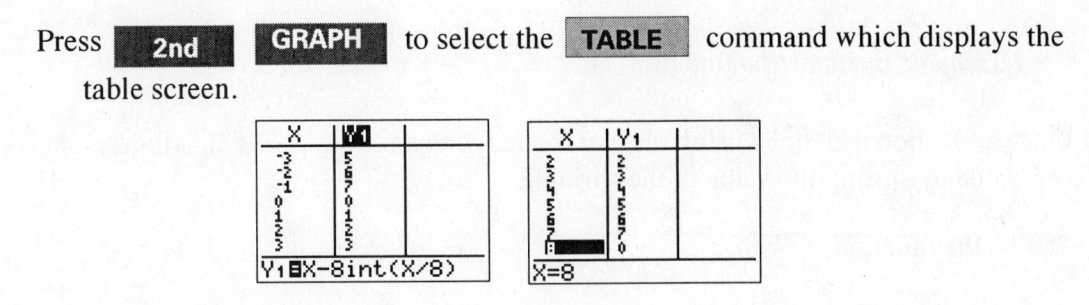

Press [2nd] [GRAPH] to select the [TABLE] command which displays the table screen.

By using the up and down arrows, you can scroll through the table, either up or down.

Storing a value for a variable

In the home screen enter the input value and store it in the variable x.

Type the number you wish to store.

Press [STO ▷]. This is the key used to store values.

When pressed, an arrow pointing right appears.

Press [X, T, Θ, n]. This will display the variable x.

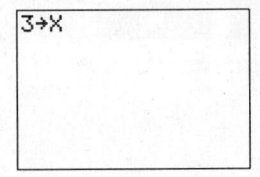

Press [ENTER]. this causes the storage to take place.

A common mistake is to forget to press [ENTER]. Store the value of 3 in x.

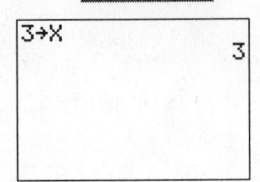

Function evaluation: storing input

Example: Evaluate the function $y(x) = -x^2 + 8x + 19$ for $x = 3$.

Once a function is defined using the **Y =** menu, you can evaluate the function for a given input value by storing the value in the variable x.

Define the function:

Press [Y=] to access the function entry screen.

Type the expression $-x^2 + 8x + 19$ for the function after $Y_1 =$.
This defines the function as Y_1.

Return to the Home Screen:

Return to the Home Screen by pressing [2nd] [MODE].

Enter the **input value** followed by [STO ▷] [X, T, Θ, n] [ENTER].
This will have the effect of storing a value for variable x.

Writing more than one command on the same line (chaining commands):

Enter a colon by pressing [ALPHA] [.]. The colon is used to chain two or more commands together on the same line.

Select the function you wish evaluated:

Select Y_1 by pressing [VARS] [▷] [1] [1].

Evaluate the function:

Press [ENTER]. The function value is displayed.

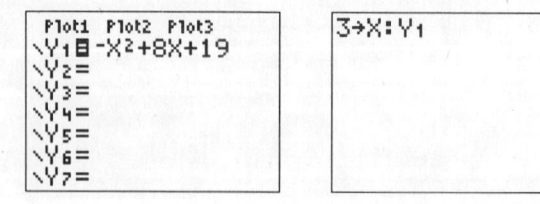

```
Plot1 Plot2 Plot3
\Y1☐-X²+8X+19
\Y2=
\Y3=
\Y4=
\Y5=
\Y6=
\Y7=
```

```
3→X:Y1
           34
```

Function evaluation using function notation

To evaluate a function using function notation you must first define the function in the Y = menu as one of the Y variables. Then on the Home Screen enter the Y variable with the input value in parentheses next to that Y variable.

Example: Evaluate the function $y(x) = -x^2 + 8x + 19$ at $x = 3$.

First, define the function:

Press **Y=** to access the function entry screen.

Type the expression $-x^2 + 8x + 19$ for the function after $Y_1 =$. This assumes that Y_1 is clear. This defines the function as Y_1.

Second, return to the Home Screen:

Return to the home screen by pressing **2nd** **MODE** .

Third, select the Y variable you wish evaluated:

Type **VARS** **▷** **1** **1** to access Y_1 in the Y variables menu. Y_1 appears in the home screen.

Finally, complete the process by giving the input value in parentheses:

Immediately after Y_1,

Type (3) followed by **ENTER** .

This returns the output of the function stored in Y_1 at an input value of 3.

```
Plot1 Plot2 Plot3
\Y1■-X²+8X+19
\Y2=
\Y3=
\Y4=
\Y5=
\Y6=
\Y7=
```

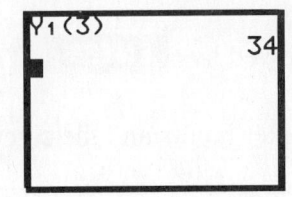

```
Y1(3)
              34
```

Function evaluation using Ask mode with the table screen

You can define a function and, using the table with the independent variable set to Ask mode and the dependent variable set to Auto mode, enter values for the independent variable and by pressing ENTER after each entry, displaying the output of the function for the input entered.

Evaluate the function $y(x) = -x^2 + 8x + 19$ at $x = 3$.

Define the function:

Press **Y=** to access the function entry screen.

Type the expression $-x^2 + 8x + 19$ for the function after $Y_1 =$. This assumes Y_1 is clear. This defines the function as Y_1.

Select the table setup **TblSet** :

Press **2nd** **WINDOW** to access **TblSet** , the table set up screen.

Arrow down to **Indpnt** and over to **Ask**.

Press **ENTER** to select the **Ask** mode for the independent variable. The dependent variable should still be in **Auto** mode.

Press **2nd** **GRAPH** to access **TABLE** , and display the table screen. A blank table appears.

Enter the input value:

Type the input followed by **ENTER** . The output for this input appears.

You may continue to enter inputs and the calculator will return the corresponding outputs.

Graphing a function

Once a function is defined in the Y = menu, its graph can be displayed on the graphics screen. It may be necessary to adjust the viewing rectangle which controls the portion of the graph shown.

Example: Graph the function $f(x) = -x^2 + 8x + 19$.

Define the function:

Press **Y=** to access the function entry screen.

Type the expression $-x^2 + 8x + 19$ for the function after $Y_1 =$.
This defines the function as Y_1.

Choose a viewing window:

A *viewing window* is a rectangular portion of the coordinate plane.

Xmin puts the left edge of the viewing window at this x-coordinate.

Xmax puts the right edge of the viewing window at this x-coordinate.

Xscl defines the distance between tick marks on the horizontal axis.

Ymin puts the lower edge of the viewing window at this y-coordinate.

Ymax puts the upper edge of the viewing window to this y-coordinate.

Yscl defines the distance between tick marks on the vertical axis.

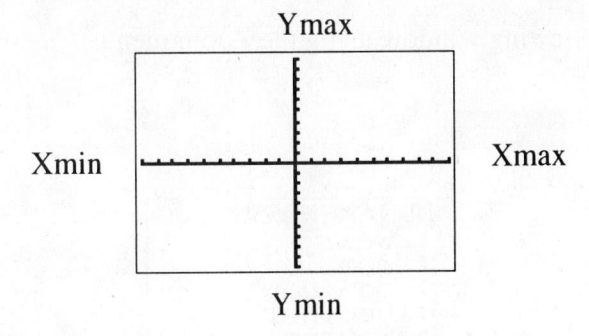

Standard viewing window (factory setting and item 6 in the Zoom menu)

The *standard viewing window* is defined as:

Xmin = –10   Ymin = –10

Xmax = 10   Ymax = 10

Xscl = 1   Yscl = 1

10 (Ymax)

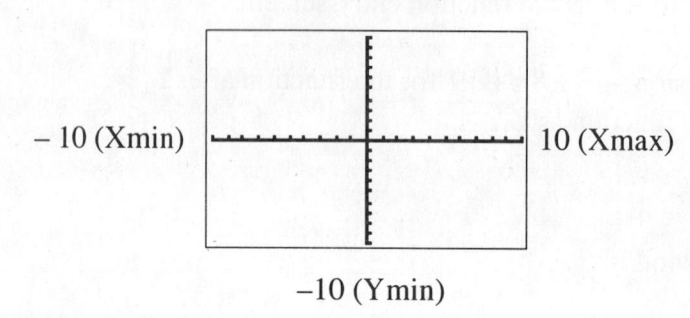

– 10 (Xmin)   10 (Xmax)

–10 (Ymin)

Set the viewing window:

Press WINDOW and enter the viewing window values.

Press the GRAPH key to display the graphics screen with a portion of the graph shown.

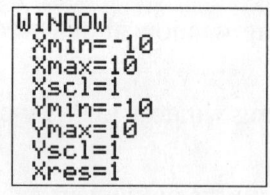

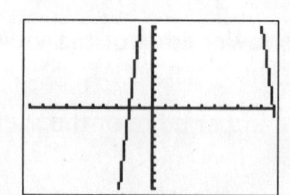

```
WINDOW
  Xmin=-10
  Xmax=10
  Xscl=1
  Ymin=-10
  Ymax=10
  Yscl=1
  Xres=1
```

To select the standard viewing window using the Zoom menu:

Press ZOOM 6 .

```
ZOOM  MEMORY
1:ZBox
2:Zoom In
3:Zoom Out
4:ZDecimal
5:ZSquare
6:ZStandard
7↓ZTrig
```

Changing the viewing window

The previous graph is only a portion of the complete graph. To see more of the graph, change the viewing window. This is done by pressing the **WINDOW** key which displays the window screen. First decide how to change the window. The table feature helps determine a better viewing window by displaying output values of the function.

Press **2nd** **GRAPH** to display the table.

Determine the smallest and largest output value between Xmin (–10) and Xmax (10).

Examine the table values for x = –10 and x = 10.

The smallest output value shown is –161. The largest output value shown is 35.

Press **WINDOW**.

Enter a value for Ymin which is smaller than the smallest output –161. –170 would be a good choice.

Enter a value for Ymax which is larger than the largest output value 35. Say, 40.

For Yscl, choose a reasonable value for the distance between tick marks. Approximately one-tenth of the distance between Ymin and Ymax is a reasonable choice. Set Yscl = 20.

Graph the function:

Press **GRAPH**.

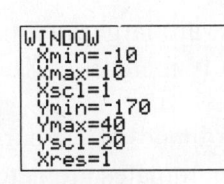

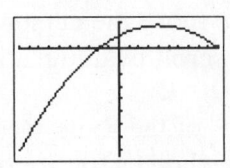

Adjusting the viewing window:

Change Xmin and Xmax to see more of the graph to the left or to the right. For instance one might increase the Xmax value to 20. It would then be wise to increase the Xscl value so there won't be an excessive number of tick marks on the x-axis.

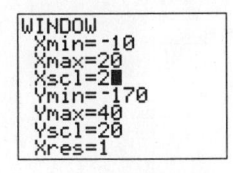

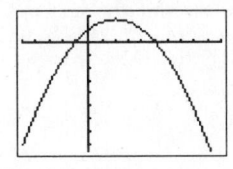

Trace: displaying coordinates of points on a graph

The cursor can be moved from one plotted point to the next along the graph of a function. The coordinates of the point located by the cursor are displayed at the bottom of the screen.

Example: Trace along the graph of $f(x) = -x^2 + 8x + 19$.

Define the function:

Enter the expression $f(x) = -x^2 + 8x + 19$ as Y_1 in the Y = menu.

Verify the viewing window settings:

If the viewing window is not in standard setting, enter the following:

Xmin = –10	Ymin = –170
Xmax = 10	Ymax = 40
Xscl = 1	Yscl = 20

Graph the function:

Press **GRAPH**. The graph of the function is displayed.

Press **TRACE**.

A flashing cursor appears on the graph along with the x and y-coordinates of the point at the bottom of the screen.

Press the left arrow key to move the cursor toward smaller inputs. Moving to the left of Xmin causes the graph to scroll by automatically decreasing Xmin.

Press the right arrow key to move the cursor toward larger inputs. Moving to the right of Xmax causes the graph to scroll by automatically increasing Xmax.

As the cursor moves, the values of the *x* and *y*-coordinates at the cursor appear at the bottom of the screen. Note that the choices for the *x*-coordinates are determined by the programming of the graphing utility. The coordinates which are displayed can be controlled as is explained in the next section of this manual.

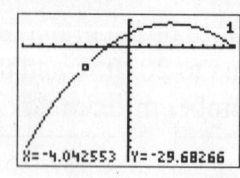

Friendly viewing windows to control the Trace Display

A "friendly" viewing window displays values for x with at most one decimal position. There are 95 pixels across the screen. The distance between Xmin and Xmax is divided by 94 to determine the size of the increment in x when moving the cursor in the graphics window.

To obtain a "friendly" window, Xmin and Xmax are selected so that when the difference Xmax − Xmin is divided by 94, the quotient terminates by the tenths position.

Tracing by x increments of 0.1: Zoom Menu Item Number 4:ZDecimal

Press **ZOOM** **4** to automatically set the "friendly" viewing window to

Xmin = −4.7   Ymin = −3.1

Xmax = 4.7   Ymax = 3.1

Xscl = 1   Yscl = 10

Notice that Xmax − Xmin = 9.4.

Divide this difference by 94, to determine the x-increment displayed. The quotient is 0.1.

The x values displayed by the Trace feature will increment by 0.1 in this window.

Using the ZOOM options to adjust the viewing window

Use the **ZOOM** key to open the following menu of options for altering the viewing window.

1:ZBox: creates a new viewing window based on a rectangle that you draw.

Press **ZOOM** **1** to return to the graphics window.

Use the arrow keys to move the cursor to one corner of the desired viewing window.

Press **ENTER** to set one corner.

Move the cursor to the opposite corner of the desired viewing window.

Press **ENTER** and the box (rectangle) becomes the new viewing window.

2:Zoom In: Use this item to zoom in using a preset factor to decrease the size of the viewing window. The factors can be set by selecting the Memory submenu from the Zoom screen and selecting item number **4:SetFactors** in that submenu.

Press `ZOOM` `2` which returns you to the graphics window.

Use the arrow keys to position the cursor at the desired center of the new window.

Press `ENTER` .

You can now repeatedly zoom in by moving the cursor to the center of the desired window and pressing `ENTER` .

3:Zoom Out: Use this item to zoom out using a preset factor to increase the size of the viewing window. The factors can be set by selecting the Memory submenu from the Zoom screen and selecting item number **4:SetFactors** in that submenu.

Press `ZOOM` `3` which returns you to the graphics window.

Use arrow keys to position the cursor at the desired center of the new window.

Press `ENTER` .

You can now repeatedly zoom out by moving the cursor to the center of the desired window and pressing `ENTER` .

4:Decimal: This item sets a viewing window so that the *x* values will increment by 0.1 as the cursor moves across the window or when using the trace feature. These are "friendly" values for analyzing the graph.

Press `ZOOM` `4` to change the window so that Xmin = –4.7, Xmax = 4.7, Ymin = –3.1, and Ymax = 3.1. Both Xscl and Yscl are 1.

5:Square: sets the viewing window so that the unit used on the x and y axes are the same distance. Graphs, such as circles, will appear distorted if this aspect ratio is not one.

Press `ZOOM` `5` .

6:Standard: returns graph to the standard viewing window.

Press `ZOOM` `6` to graph in the standard viewing window.

7:Trig: sets a "friendly" viewing window for graphing trigonometric functions.

Press `ZOOM` `7` to select this window.

8:Integer: sets the viewing window so that the x values will increment by one as the cursor moves across the window. These are "friendly" values for analyzing graphs.

Press ZOOM 8 to change the window.

9:ZoomStat: This item resets the viewing window so that all the points in the active StatPlots are displayed.

Press ZOOM 9 to change the window.

0:ZoomFit: This item resets the Ymin and Ymax to include the minimum and maximum values of the selected functions between the current Xmin and Xmax and graphs the selected functions in this new viewing window. Xmin and Xmax are not changed.

Press ZOOM 0 to change the window.

Connected and Dot Modes for graphs

The TI-83 has a graphics mode setting icon for each function in the Y= menu. The icon appears to the left of the y variable and may be changed by placing the cursor on the icon and pressing the enter key to cycle through the seven choices. For **Dot** mode the icon should appear as three separate dots. Connected mode is the factory setting and is four dots close together.

As an example the graph of the function $y(x) = x^2 - 1$ is shown with both mode settings using the "friendly" viewing window set by choosing item number **4:ZDecimal** in the **Zoom** menu.

Dot mode

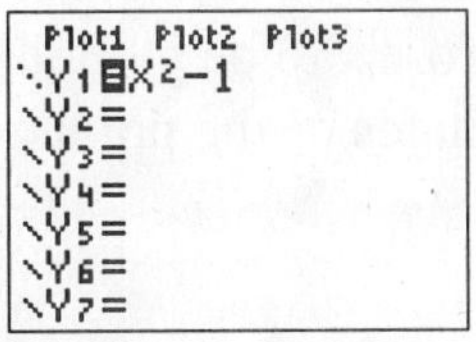

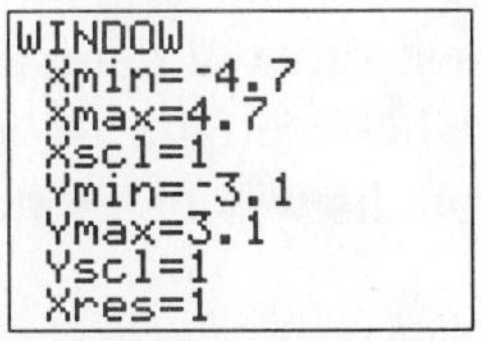

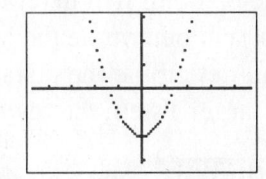

Connected mode

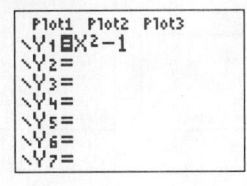

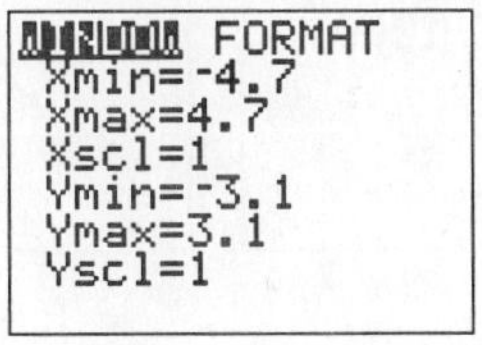

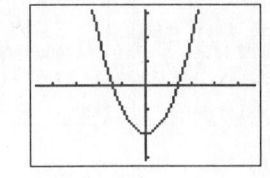

245

Parametric graphing

Parametric equations consist of two equations, representing the *x*-component and the *y*-component of a graph, each expressed in terms of a third independent variable, *T*. Parametric equation may be used to control the points which are plotted on the graph of a function when **Dot** mode is selected.

Before entering the two parametric equations, you must change from function mode to parametric mode and adjust your viewing window.

Press MODE and use the ▽ key to highlight **Par** on the mode screen.

Press ENTER.

Example: Suppose we wish to plot the function $y(x) = 10x - 1000$ but only show the points on the graph whose *x*-coordinates are 0, 5, 10,... 200. First we select Parametric in the Mode window.

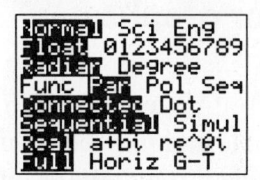

Now we open the Y= menu and define $x = t$ and $y = 10t - 1000$ and select dot mode using the mode icon next to the y variable.

Press Y= and enter *t* for X_{1T} and $10t - 1000$ for Y_{1T} and select the dot icon next to X_{1T}.

Press WINDOW and set the viewing window appropriately. (Shown below.)

The value of Tmin will determine the smallest *t* and hence x-coordinate to be plotted and the Tmax will determine the largest value of *t* and therefore *x* to be plotted. The size of the Tstep will determine the distance between the *x*-coordinates of the points which are plotted. The larger Tstep, the fewer points will be plotted.

Press GRAPH when all this is set up.

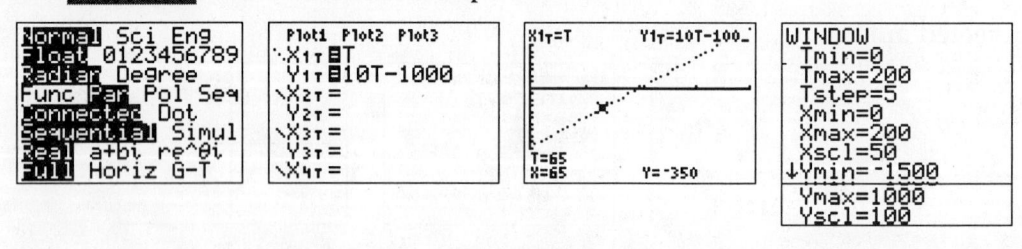

When you use the trace feature you will see the points on the graph highlighted by the cursor and the *t*, *x* and *y*- coordinates will be displayed at the bottom of the screen. As you move the cursor only those points whose *x*-coordinates are multiples of 5 starting with 0 are displayed.

Graphing the solution interval to inequalities in one variable

Inequalities in one variable, such as $2x + 3 < 5$, can be entered directly in the Y= menu. The calculator will graph the solution set by drawing a horizontal line at $y = 1$ for all x values that make the inequality true. The "1" is the calculator's way of saying "true" whereas "0" is what the calculator uses for "false".

Set the calculator to **Func** mode:

Press MODE , arrow down to **Func**, and press ENTER .

Press Y= .

Type the left hand side of the inequality.

Open the **TEST** menu by pressing 2nd MATH .

Select the desired inequality relation by typing the number next to the relation, in this case **5**. The calculator returns to the Y= screen and prints the inequality symbol.

Type the right hand side of the inequality.

Set an appropriate viewing window. It is important that $y = 0$ and $y = 1$ be clearly visible on the graphics screen. It would be appropriate to use Ymin = –2 and Ymax = 2.

Press GRAPH . A horizontal line segment at $y = 1$ appears for x values that satisfy the inequality.

You can trace on the graph. An output of 1 indicates the corresponding x is in the solution set to the inequality. An output of 0 indicates the corresponding x makes the inequality false.

Example: Graph the solution set to $2x + 3 < 5$.

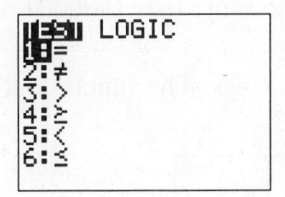

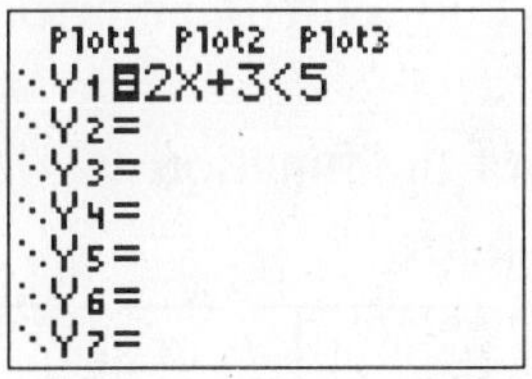

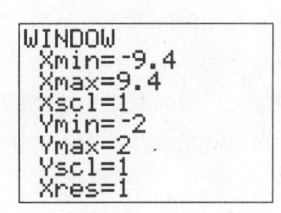

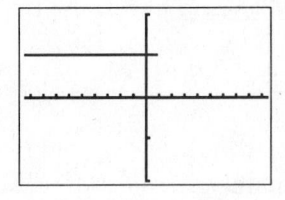

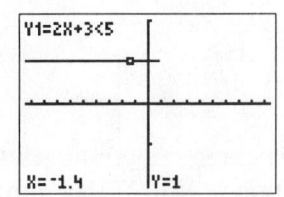

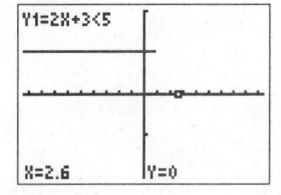

Finding zeros using the CALCULATE menu item 2: zero

Enter the function in the Y= menu(the mode setting should be **Func**) and graph the function in a window so that the points where the graph crosses the horizontal axis are visible in the graphics screen.

Press `2nd` `TRACE` to display the **CALCULATE** menu, abbreviated as CALC in yellow above the Trace key.

Type `2` to select **2:zero** from the menu.

The calculator returns to the graphics window and prompts you to select a **Left Bound** for one of the roots.

Use the **left arrow key** to move the cursor until the value for x at the cursor is smaller than the desired zero.

Press `ENTER` and the x value of the cursor is selected as the left bound.

The calculator prompts you to select a **Right Bound** for the zero.

Press the **right arrow key** at least twice to move the cursor until the value for x at the cursor is larger than the desired zero.

Press `ENTER` and the x value of the cursor is selected as the right bound.

The calculator prompts you to select a **Guess** for the zero.

Press the **left arrow key** to move the cursor close to the zero, which is where the graph intersects the x-axis. Be careful not to move the cursor past the left bound. The guess must be between the left and right bounds. When the cursor is close to the zero, press `ENTER`.

The calculator begins approximating the zero. A moving vertical line segment is displayed in the upper right corner of the screen to indicate that the calculator is computing. When finished, the calculator displays the approximate zero within the interval defined by the left and right bounds. This solution is correct with an error of no more than 0.00001.

Example: Find the largest zero of the function $y = x^2 + x - 3$. The function is graphed in the standard viewing window.

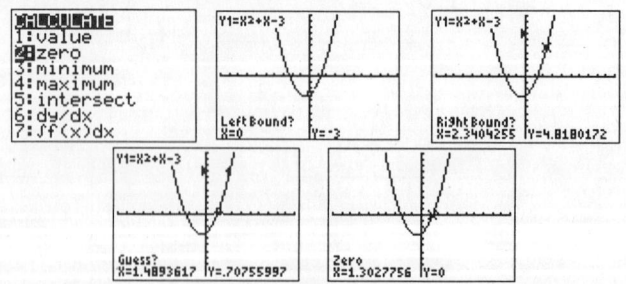

The largest zero or root of the function is 1.3027756, approximately.

Finding intersection points using the CALCULATE menu

Enter the two functions in the Y= menu as Y_1 and Y_2. (You may use any of the other y variables if you do not wish to erase functions you may have entered for Y_1 and Y_2.)

Graph both functions in a window so that the desired intersection point is visible on the graphics screen.

Press **2nd** **TRACE** to display the **CALCULATE** menu.

Press **5** to select **5:intersect**.

The calculator returns to the graphics screen. The calculator prompts you to select the **First curve**, meaning the graph of one of the functions. The cursor is on the graph of Y_1.

Press **ENTER** to select Y_1 as the first curve.

Now the calculator prompts you to select the **Second curve**. The cursor is on the graph of Y_2. (If there were more than two graphs on the graphics screen you could move the cursor to one of the other graphs by using the up or down cursor control.)

Press **ENTER** to select Y_2 as the second curve.

The calculator prompts you to supply a **Guess** for the intersection point.

Do this by using the right or left arrow keys to move the cursor near the intersection point.

Press **ENTER** when the cursor is close to the intersection point.

The calculator begins approximating the intersection point. A moving vertical line segment is displayed in the upper right corner of the screen to indicate that the calculator is computing. When finished, the calculator displays the approximate ordered pair for the intersection point. The solution is correct with an error of no more than 0.00001.

Example: Find the intersection point of the functions $y = 2x - 1$ and $y = -3x + 7$. The graphs appears in the standard viewing window.

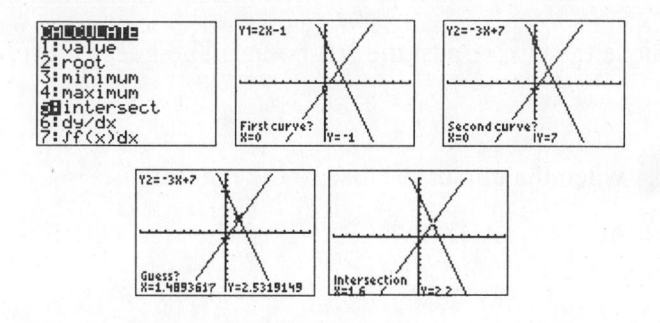

The intersection point is (1.6, 2.2).

Finding a function's maximum using the CALCULATE menu

Enter the function in the Y= menu.

Graph the function in a window so that the graph displays a highest point.

Press **2nd** **TRACE** to open the **CALCULATE** menu.

Type **4** to select **4:maximum**.

The calculator returns to the graphics window. The calculator prompts the user to select a **Left Bound** for the interval containing the function's maximum.

Use the **left arrow key.** Move the cursor until it is to the left of the function's maximum.

Press **ENTER** to select this x value as the left bound.

Now the calculator prompts the user to select an **Right Bound** for the interval containing the function's maximum.

Press the **right arrow key** at least twice to move the cursor until the cursor is to the right of the function's maximum.

Press **ENTER** to select this x value as the right bound.

The calculator prompts the user to select a **Guess** for the function's maximum.

Press the **left arrow key** to move the cursor close to the function's maximum.

Be careful not to move the cursor past the left bound. The guess must be between the left and right bounds.

Press **ENTER** when the cursor is close to the maximum.

The calculator begins approximating the function's maximum. A moving vertical line segment is displayed in the upper right corner of the screen to indicate that the calculator is computing. When finished, the calculator displays the approximate maximum of the function within the interval defined by the left and right bounds. This solution is correct with an error of no more than 0.00001.

Example: Find the vertex of $y = -x^2 + x + 4$. For this function the vertex is the highest point on the graph. The graph appears in the standard viewing window.

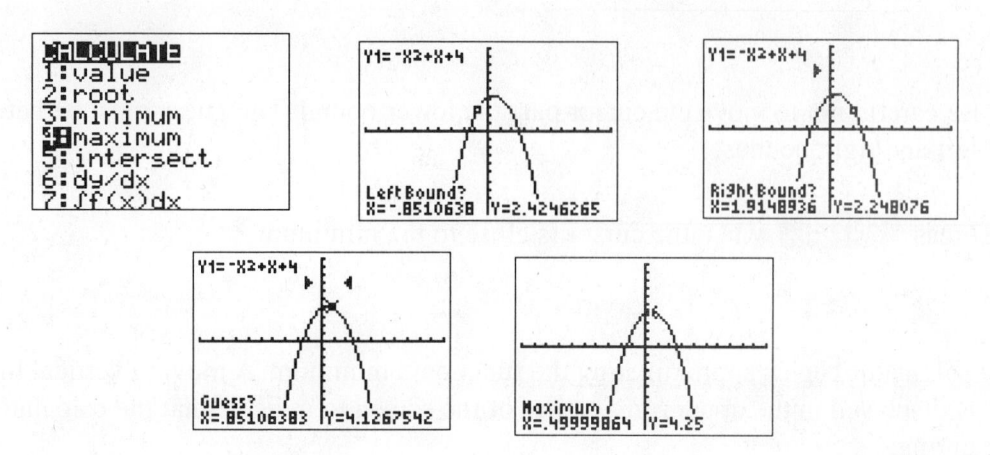

The vertex is actually the point (0.5, 4.25). The x-coordinate shown is an approximation.

Finding a function's minimum using CALC

Enter the function in the Y= menu.

Graph the function in a window so that the graph displays a lowest point.

Press **2nd** **TRACE** to open the **CALCULATE** menu.

Type **3** to select **3:minimum**.

The calculator returns to the graphics window. The calculator waits for you to select a **Left Bound** for the interval containing the function's minimum.

Use the **left arrow key** to move the cursor until it is to the left of the function's minimum.

Press **ENTER** and this *x* value becomes the left bound.

The calculator waits for you to select a **Right Bound** for the interval containing the function's minimum.

Press the **right arrow key** at least twice to move the cursor until it is to the right of the function's minimum.

Press ENTER and this *x* value becomes the right bound.

The calculator waits for you to select a **Guess** for the function's minimum.

Press the **left arrow key** to move the cursor close to the function's minimum.

Be careful not to move the cursor past the lower bound. The guess must be between the left and right bounds.

Press ENTER when the cursor is close to the minimum.

The calculator begins approximating the function's minimum. A moving vertical line segment is displayed in the upper right corner of the screen to indicate that the calculator is computing.

When finished, the calculator displays the approximate minimum of the function within the interval defined by the left and right bounds.

This solution is correct with an error of no more than 0.00001.

Example: Find the vertex of $y = x^2 - 3x - 2$. The vertex of the graph of this function is the lowest point on the graph. The graph appears in the standard viewing window.

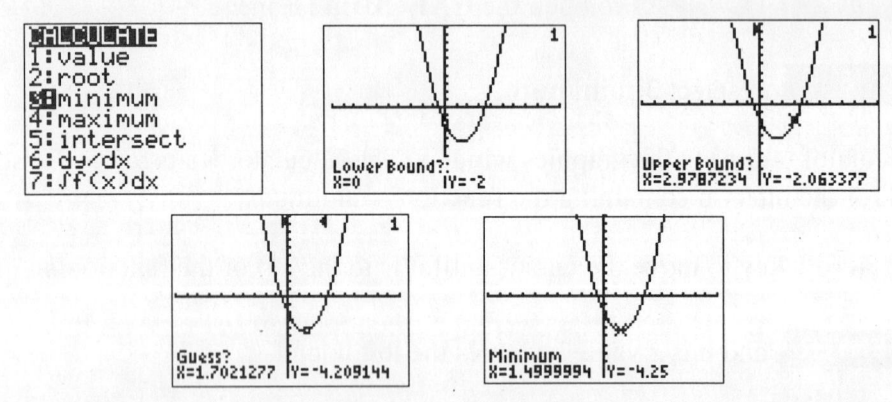

The vertex is actually (1.5, −4.25). The x-coordinate shown is only an approximation.

Working With Lists

Entering data into a list

A list of data can be entered, displayed, copied to another list, stored, sorted, used to graph families of curves, and used in mathematical expressions. Data that has been entered in lists can be displayed graphically using the **Stat Plots** menu. The types of plots available include scatter plots, line plots, box-and-whisker plots, and histograms.

From the Home Screen:

Press `STAT` `ENTER` to select **1:Edit** form the **STATEDIT** menu and to display the list screen.

With the lists displayed:

Position the cursor in the column with heading L_1 using the arrow keys. At the bottom of the screen you should see $L_1(1) =$. The 1 in the parentheses indicates that the cursor is at the first position in the list.

Enter each of the data elements.

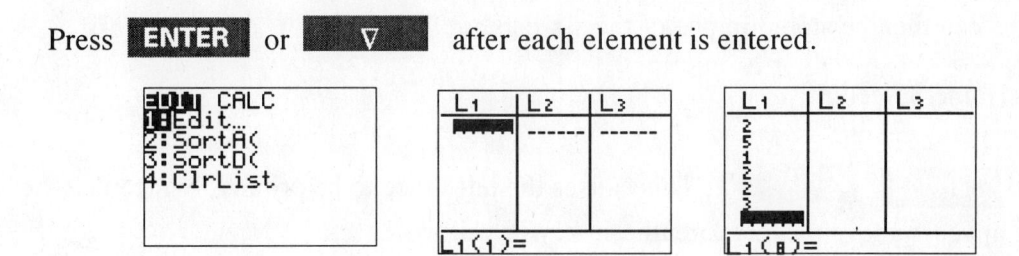

Press `ENTER` or `▽` after each element is entered.

Clearing data from a list

If data is already entered in a list you wish to use and you would like to erase the entire list:

Use the `△`, `▷` or `◁` keys and position the cursor in the column heading over the list name.

Press `CLEAR` `ENTER` to remove the existing data elements and clear the list.

The empty list is ready for entry of new data elements.

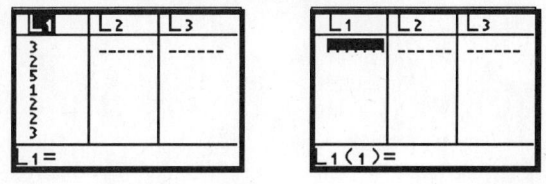

Changing numbers in an existing list

A number in a list can also be changed by positioning the cursor over that number and typing in the desired new number, followed by **ENTER** .

Inserting new numbers into an existing list

If you wish to insert a new number into an existing list without removing any of the numbers from the list first position the cursor where you wish to place the new number. Then put the calculator in **insert** mode by pressing **2nd** **DEL** . The abbreviation for **insert** appears above the **DEL** key in yellow. A zero appears in the list where you wish to insert a new number. Type the new number and press the enter key.

Entering data into a list from the Home Screen

Data entered on the Home Screen using list notation and with each element followed by a comma can then be stored in one of the six lists.

From the Home Screen:

Press **2nd** **(** . This causes the left brace to be printed. Notice that the left brace appears above the left parenthesis key in yellow.

Enter the data elements separating each element with a comma.

Press **2nd** **)** after the last data element is entered. This prints the right brace which appears above the right parenthesis key in yellow.

```
{                    {1,2,3,4,5}
```

Storing and viewing the newly-generated list:

After entering the list in braces on the home screen press **STO ▷** **2nd** **1** , then press **ENTER** to store the list in L_1. Notice that L_1 appears above the key with 1 on it, printed in yellow.

Press **STAT** **ENTER** to select **1:Edit** from the **STATEDIT** menu to view the newly-generated list on the list screen.

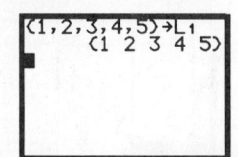

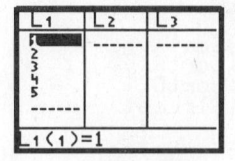

Pressing **STO** **2nd** **2** , and then **ENTER** would store the list in L_2.

Press **STAT** **ENTER** to select **1:Edit** from the **STATEDIT** menu and view the newly-generated list on the list screen.

Generating new data on the list screen:

Press **STAT** **ENTER** to select **1:Edit** from the **STATEDIT** menu and to display L_1, L_2, and L_3.

Position the cursor on the column heading L_2 using the arrow keys.

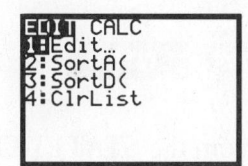

Create a new list L_2, by adding three to each element of L_1:

Type $L_1 + 3$ and press **ENTER** . The calculator adds 3 to each element of L_1 and stores this new list in L_2.

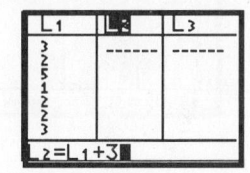

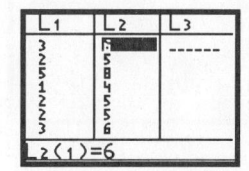

Applying arithmetic operations to an existing list to create new list

A new list can be created either on the home screen and then stored in a blank list or on the list screen.

Generating new data from the Home Screen

A new list of data can be generated from the home screen using an existing list name. The data generated can be stored as a new list and viewed on the list screen.

From the Home Screen:

Press **STAT** **ENTER** to select the **EDIT** menu and to display L_1, L_2, and L_3.

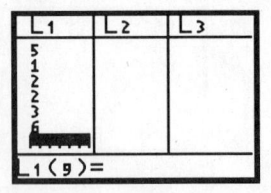

Create a new list L2, adding three to each element of an existing list L1:

Press **2nd** **QUIT** to return to the Home Screen.

Type $L_1 + 3$

Store and view the newly-generated list into L_2.

Press **STO** **2nd** **2**, then press **ENTER** to store the list in L_2.

Press **STAT** **ENTER** to select **1:Edit** from the **STATEDIT** menu and view the newly-generated list.

Use right and left arrows to view additional elements of the list on the Home Screen.

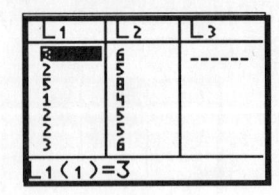

256

Generating a list of data using the sequence function in the LIST OPS menu

From the Home Screen access the **LIST OPS** menu:

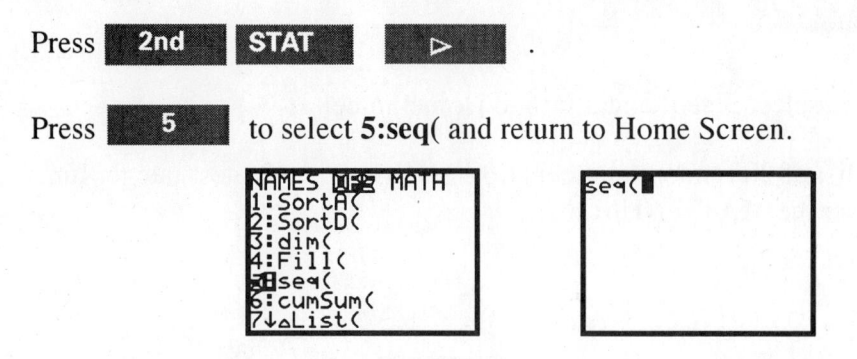

Press `2nd` `STAT` `▷` .

Press `5` to select **5:seq(** and return to Home Screen.

Type $x^2, x,$ **1, 10, 1**) to generate the squares of the numbers 1–10.

Press `ENTER` to display the list of elements on the Home Screen. Use left and right arrows to view additional list elements on the Home Screen.

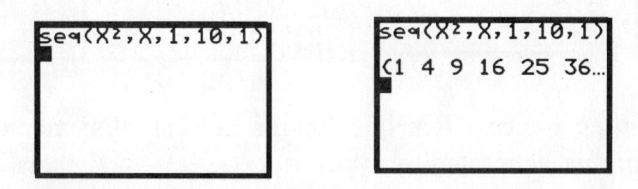

Note: The general format for entry of a sequence is:

> seq (generating rule, variable used in generating rule, initial value of the variable,
> ending value of the variable, increment of the variable)

The increment of the function may be left out on the TI-83 if you wish the increment to be the default value of 1. Commas are necessary and must be included after each input for the sequence function.

To store and view the newly-generated list in L_1:

Press `STO` `2nd` `1` , then press `ENTER` to store the list in L_1.

Press `STAT` `ENTER` to return to the list screen.

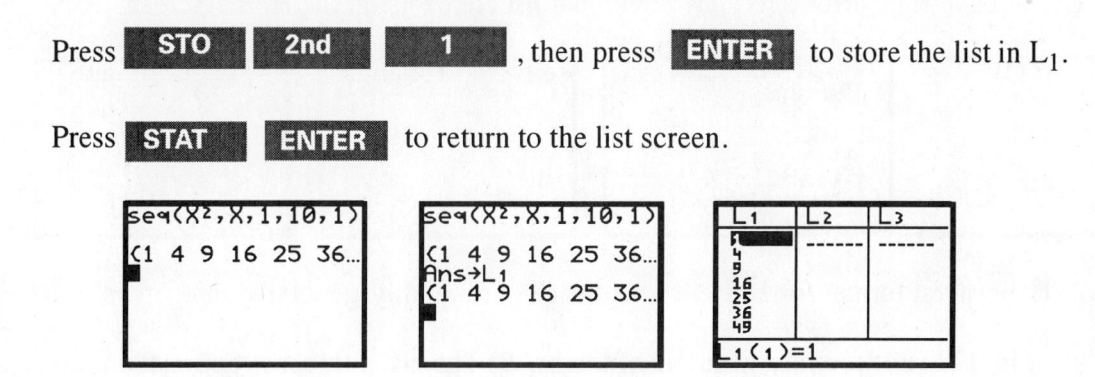

Generating a list of random numbers using sequences

From the Home Screen access the **LIST OPS** menu:

Press `2nd` `STAT` `▷` .

Press `5` to select **5:seq(** and return to Home Screen.

Generate and store a list of 20 random numbers (0 – 9) using the greatest integer function, **int**, accessed from the **MATH NUM** menu.

Press `MATH` .

Use the right arrow to highlight **NUM**. Enter `5` to select the greatest integer function *5: int*.

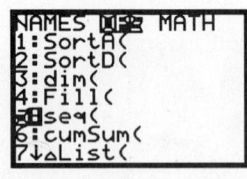

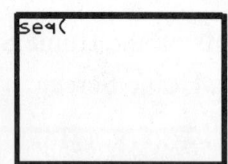

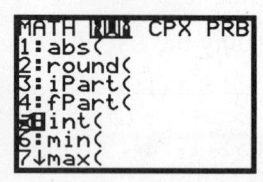

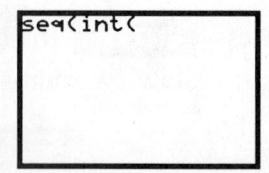

To generate a random integer access **RAND** from the **MATH PRB** menu and multiply the random decimal number generated by 10:

Enter (**10** and Press `MATH` .

Use the right or left arrow to highlight **PRB.**

Press `ENTER` to select the random number generator.

Complete the sequence by typing), *x*, **1, 20, 1**). The screen should look like the one below.

Press `ENTER` to display the list of elements on the Home Screen .

Use left and right arrows to view additional list elements on the Home Screen.

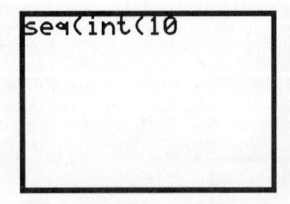

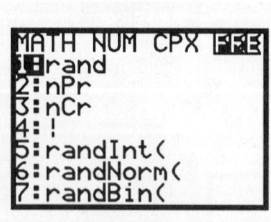

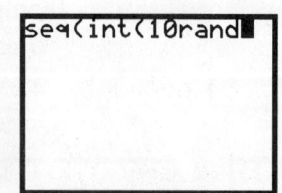

Note: The general format for entry of a sequence of randomly-generated integers is:

seq (int (10 rand), *x,* **1, number of terms, 1)** The last number is optional on the TI-83.

258

Storing the newly-generated list in L_1:

Press **STO** **2nd** **1** , then press **ENTER** to store the list in L_1.

Press **STAT** **ENTER** to view the list on the list screen.

Use up and down arrows to view additional list elements on the list screen.

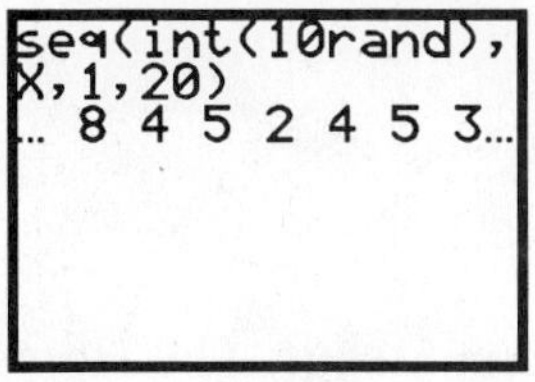

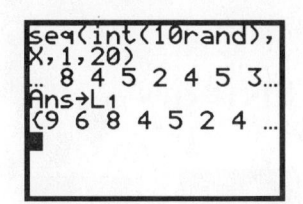

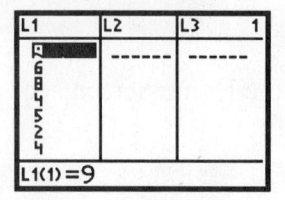

Adding a random number to each element of a list:

With the lists screen displayed:

Position the cursor on column heading L_2. At the bottom of the screen is displayed $L_2 =$

Complete the right-hand side by entering L_1 + int(10 RAND).

Press **ENTER** to display the list.

Use up and down arrows to view additional list elements.

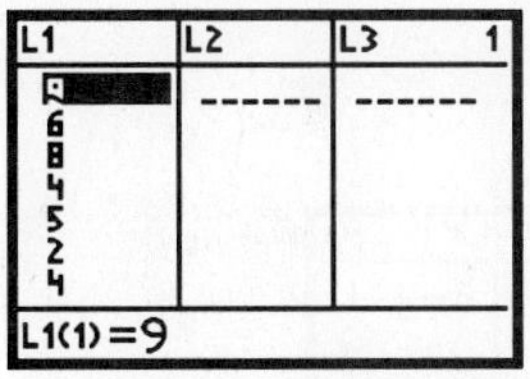

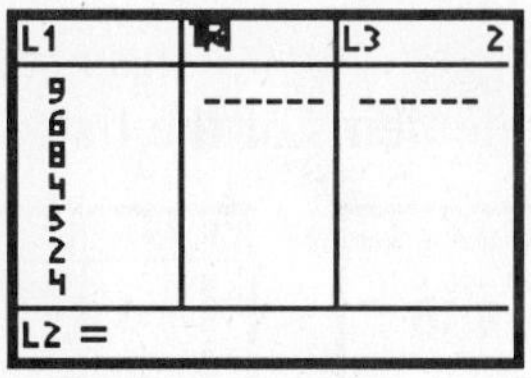

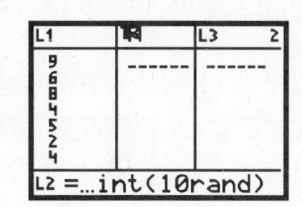

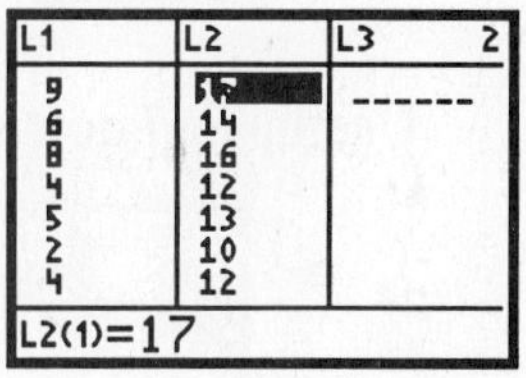

NOTES:

- **RAND** generates and returns a random number between zero and 1 (0 < rand < 1).

- To generate more random number sequences, have individuals each store a different **integer** seed value in **RAND** first.

- If using a new set of TI-82s or TI-83s, have each person store an integer (social security numbers are a good choice) to generate random numbers.

- When you reset a TI-82 or TI-83 or store **0** to **RAND**, the factory-set seed value is restored.

Generating data using a sequence and an existing list (from the Home Screen)

A new list of data can be generated using an existing list with a sequence command from the Home Screen or with the lists displayed.

Using the list of twenty numbers randomly generated, return to the Home Screen.

Press **2nd** **QUIT**.

From the Home Screen access the **LIST OPS** menu:

Press **2nd** **STAT** **▷** **5** to select **5:seq(**. The Home Screen displays **seq(**.

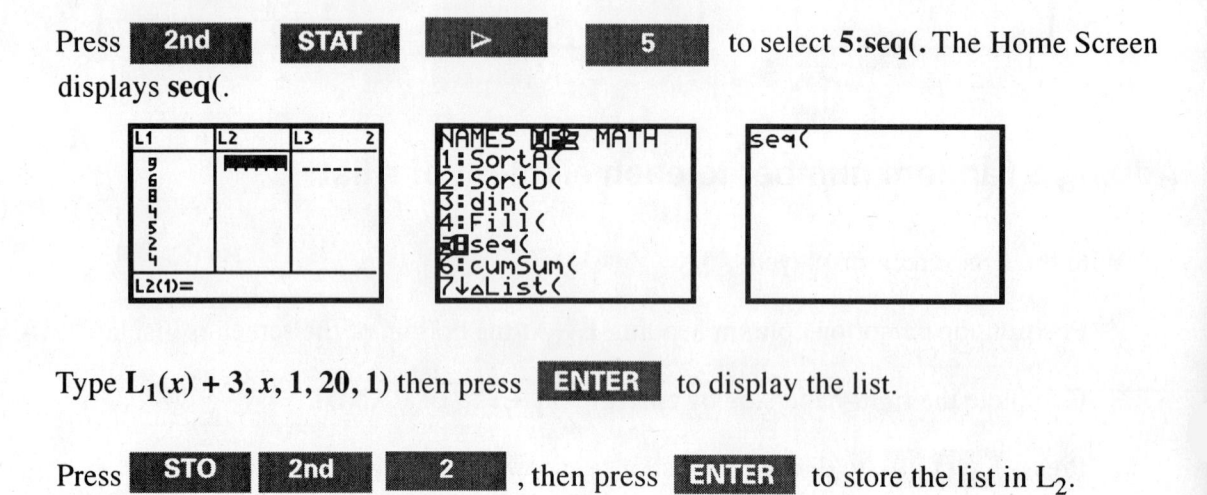

Type **L₁(x) + 3, x, 1, 20, 1)** then press **ENTER** to display the list.

Press **STO** **2nd** **2**, then press **ENTER** to store the list in L_2.

Press **STAT** **ENTER** to return to the list screen.

Use up and down arrows to view additional elements of the list.

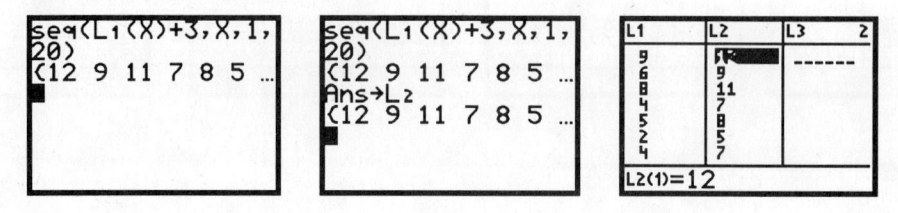

Note: The general format for entry of a sequence using an existing list is:

seq (L_{old} (x) with operation(s), x, 1, number of elements in the list,1) The last one is optional on the TI-83. It need only be included when you wish the step to be something other than 1.

Constructing a scatter plot of data

A scatter plot is a two dimensional graph of ordered pairs and requires two lists of data values, one for the first coordinates and one for the second coordinates. Scatter plots plot the data points from two lists called the Xlist and Ylist in the **SCATTER PLOTS** menu.

You can select one of three options for highlighting the points: (a) a box (□), (b) a cross (+) or (c) a dot (•). The number of data points in the Xlist and in the Ylist must be the same. The frequency of occurrence of each data value does not apply to scatter plots.

From the Home Screen:

Press **Y =** and turn off all selected Y= equations.

Press **STAT** **ENTER** to view the list screen and display L_1, L_2, and L_3.

If data has been previously entered in L_1, L_2, and L_3, clear existing data.

Enter the first set of data in L_1 and the second set of data in L_2.

Press **2nd** **Y =** to access the **STAT PLOTS** menu.

Select **1** for **Plot1**. Highlight **On** and press **ENTER**.

Use the **▽** and **▷** to highlight the first icon under the **Type:** options.

Press **ENTER**. This darkens the first icon which is the **scatter plot** icon.

Use **▽** the list under **Xlist**. Type **L_1** for the **Xlist**, and type **L_2** for the **Ylist**. Choose the square in the **Mark:** options.

Press **ENTER** after each selection.

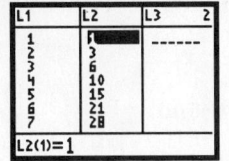

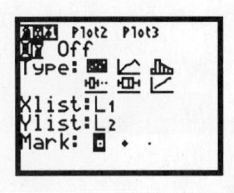

Set a viewing window that contains all the data points and then display the scatter plot on the graphics screen.

Press **ZOOM** **9** to select **9:ZoomStat** from the **ZOOM** menu, set a viewing window and display the scatter plot.

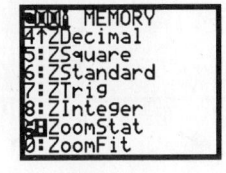

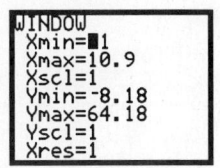

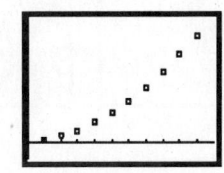

Constructing a line plot of data

Enter the data in lists L₁ and L₂.

Press **2nd** **Y=** to access the **STAT PLOTS** menu.

Select **2** for **Plot2**. Highlight **On** and press **ENTER**.

Use the **▽** and **▷** to highlight the second icon in the **Type:** options.

Press **ENTER**. This selects the second type which is the **line plot** option.

Use the **▽** and type **L₁** in **Xlist,** and **L₂** in **Ylist** if it isn't already selected.

Press **ENTER** after each selection.

Select **ZoomStat** by pressing **ZOOM** **9**. This command sets a viewing window that contains all the data points and then displays the scatter plot. Turn off **Plot1**.

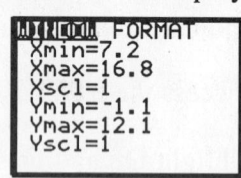

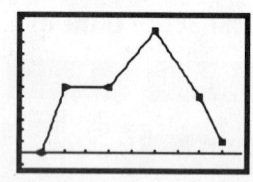

Turning off Stat Plots

When finished with them, it is a good idea to turn off the statistical plots. Not doing so often results in confusing graphs and error messages at a later time when you may be graphing functions entered in the Y= menu.

The TI-83 displays the Plot names at the top of the Y= menu. Any Plot which is darkened is turned on. To turn a plot off simply move the cursor to the Plot name and press the enter key.

Press **2nd** **Y=** to access the **STAT PLOTS** menu.

Select **4** for **PlotsOff**. The calculator returns to the home screen and displays the message **PlotsOff**.

Press **ENTER**. The message **Done** appears indicating that the statistical plots are now off.

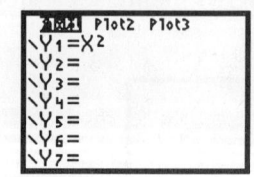

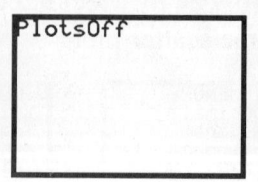

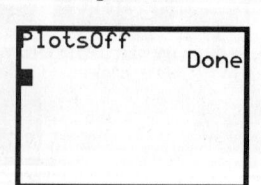

Finite differences and ratios using the sequence function

Finite differences can be calculated using the sequence function for data entered in a list or by using the **ΔList** item in the **List Ops** menu. If the data consists of ordered pairs, enter the x-coordinates in the first list and the y-coordinates in the second list.

Example: Enter the following table into two list on the list screen and calculate the finite differences for the values in the second column:

X	S(x)
1	35,000.00
2	37,000.00
3	39,000.00
4	41,000.00
5	43,000.00
6	45,000.00
7	47,000.00
8	49,000.00
9	51,000.00
10	53,000.00

If the function is known and entered in the Y= menu, then the ordered pairs can be entered into two lists element by element or by using the function rule with the sequence command. In the table above, the left column is a list of sequential numbers (years) from 1 to 10; the right column is a list of the first ten years of salaries for a fictitious company named CD-R-Us, given an initial starting salary of $35,000, with annual pay raises of $2000 each subsequent year.

Enter the column of sequential years, X, as the first list, L_1, using the sequence function:

Press **STAT** **ENTER** to view the list screen and to display L_1, L_2, and L_3.

Position the cursor on the column heading L_1 using the arrow keys.

Press **2nd** **STAT** **▷** **5** to select **5:seq(**.

Type **x , x, 1, 10)** to generate the list of numbers 1 – 10. Press **ENTER** .

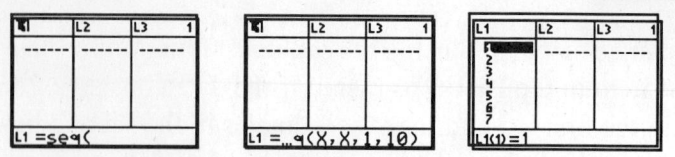

Enter the column of salaries, S(X), as the second list, L_2:

Position the cursor on the column heading L_2 using the arrow keys.

Press **2nd** **STAT** ▷ **5** to select **5:seq(**

Type: **35000 + 2000(x-1), x, 1, 10)** to generate the list of salaries. Press **ENTER** .

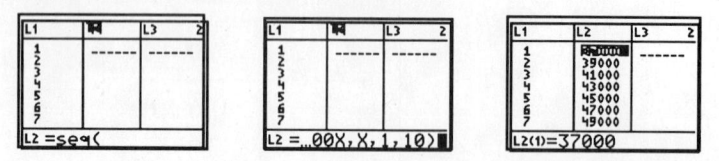

Generate the first finite differences of L_2 and store in L_3 using the sequence function:

Position the cursor on the column heading L_3 using the arrow keys.

Press **2nd** **STAT** ▷ **7** to select **7:ΔList.**

Type L_2) to generate the list of first finite differences.

Press **ENTER** .

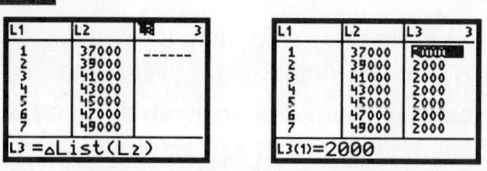

Generate the finite ratios of L_2 and store in L_4:

Position the cursor on the column heading L_4 using the arrow keys.

Press **2nd** **STAT** ▷ **5** to select **5:seq(.**

Type L_2 (x + 1) / L_1 (x), x, 1, 9) to generate the list of finite ratios.

Press **ENTER** .

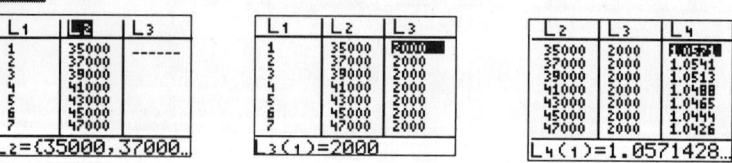

Note: The first finite difference is constant, an indication that the function is linear.

Using finite differences to select an appropriate regression equation

Data generated by an unknown sequence can be investigated using finite differences in order to select an appropriate regression equation and determine the function.

Example: Use finite differences to investigate the sequence

$$1, 3, 6, 10, 15, 21, 28, 36, 45, 55, \ldots$$

The first list consists of the positions of the known terms in the sequence; the second list consists of the terms that are given of the unknown sequence.

Position: 1, 2, 3, 4, 5, 6, 7, 8, 9, 10, ...

Seq term: 1, 3, 6, 10, 15, 21, 28, 36, 45, 55, ...

Enter the position numbers as the first list, using a sequence.

Press **STAT** **ENTER** to select the **STAT EDIT** menu and to display L_1, L_2, and L_3.

Position the cursor on the column heading L_1 using the arrow keys.

Press **2nd** **STAT** **▷** **5** to select **5:seq(**.

Type **x , x, 1, 10)** to generate the list of numbers 1 – 10.

Press **ENTER**.

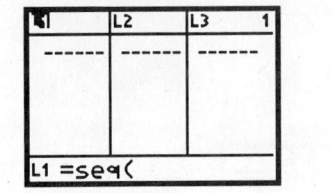

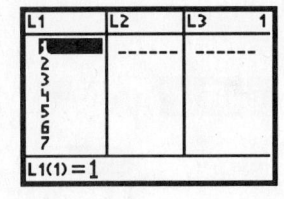

Enter the terms of the sequence as the second list, L_2, element by element.

Position the cursor in L_2 for the first entry.

Enter the terms of the sequence, element by element. Press **ENTER** after each term.

265

Generate the first finite different as L_3:

Position the cursor on the column heading L_3 using the arrow keys.

Press `2nd` `STAT` `7` to select **7:ΔList(**.

Type L_2) to generate the list of first finite differences.

Press `ENTER`.

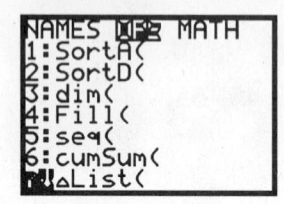

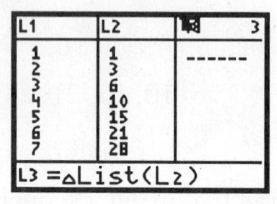

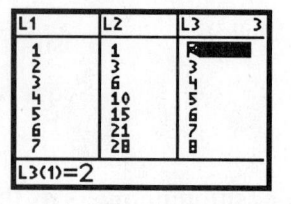

A linear model is indicated by the first finite difference being constant. If, however, the first finite differences are not constant, the function is not linear and you should continue to take finite differences.

Generate the second finite difference as L_4:

Position the cursor on the column heading L_4 using the arrow keys.

Press `2nd` `STAT` `▷` `7` to select **7:ΔList(**.

Type L_3) to generate the list of second finite differences.

Press `ENTER`.

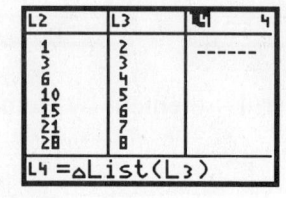

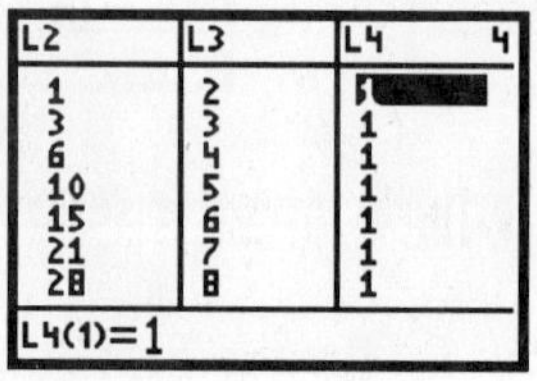

Note: The second finite difference is constant, an indication that the function is quadratic. The function can be determined using the quadratic regression model.

Creating a linear regression model

The TI–83 has the capability of creating the equation of a line that best fits a set of paired data. Such a line is called a linear regression model for the data.

Example: Enter the following paired data as lists L_1 and L_2.

x	1	2	7	9	12
y	4	11	23	32	45

Enter the paired data in two lists, L_1 and L_2.

Construct a scatter plot of the data (see index of procedures).

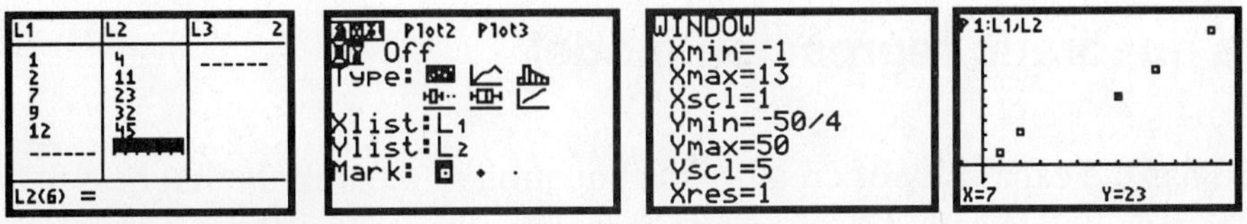

Press **STAT** ▷ to select the **STAT CALC** menu.

Press **4** **ENTER** to select **4:LinReg(ax+b)**. The calculator returns to the home screen.

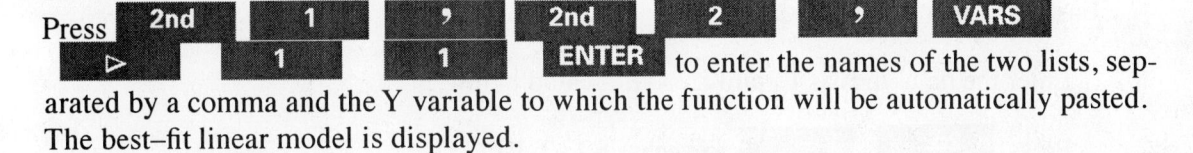

Press **2nd** **1** **,** **2nd** **2** **,** **VARS** ▷ **1** **1** **ENTER** to enter the names of the two lists, separated by a comma and the Y variable to which the function will be automatically pasted. The best–fit linear model is displayed.

The value of *a* is the slope of the line.

The value of *b* is the vertical intercept of the line.

The value of *r* measures how well the line fits the data. The closer the absolute value of *r* is to one, the better the fit. If your calculator does not display the values of r and r^2 choose the CATALOG above the zero key and scroll down to DiagnosticOn and press the enter key.

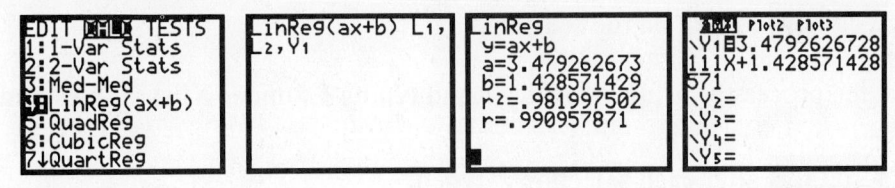

The function has been pasted into the Y= menu for Y_1.

Graph the regression model

The regression model is stored in the **Y =** menu.

Adjust the viewing window if necessary.

Press **GRAPH** to display the regression model.

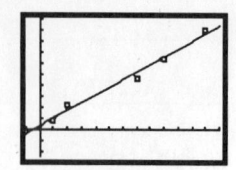

Creating a quadratic regression model

The TI–83 has the capability of creating the equation of a parabola that best fits a set of paired data. Such a line is called a quadratic regression model for the data.

The first step in creating a quadratic regression model is to construct a scatter plot of the data. A graph can be constructed using the data entered into two lists and the statistical plot capability of the calculator.

Construct a scatter plot of data

Enter the data in lists L_1 and L_2 as indicated below.

Press **2nd** **Y =** to access **STAT PLOT**.

Select **1** for **Plot1**.

Highlight **On** and press **ENTER**.

Use the arrow keys to move down to **Type** and highlight the first figure (scatter plot).

Press **ENTER** to turn on the **scatter plot**.

Repeat the process of using arrow keys and typing L_1 under **Xlist**, and L_2 under **Ylist**.

Press **ENTER** after each selection is typed.

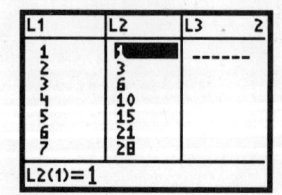

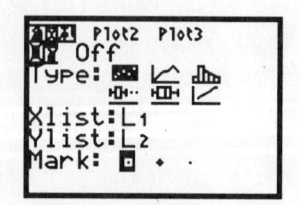

268

Set a viewing window that contains all the data points and then display the scatter plot.

Set an appropriate viewing window and graph the data.

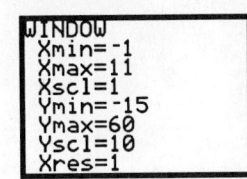

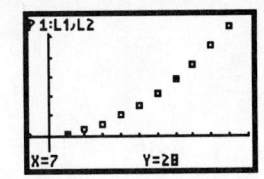

Select an appropriate regression model.

The data points displayed on the scatter plot give the appearance of being quadratic. A good first choice would be the quadratic regression model.

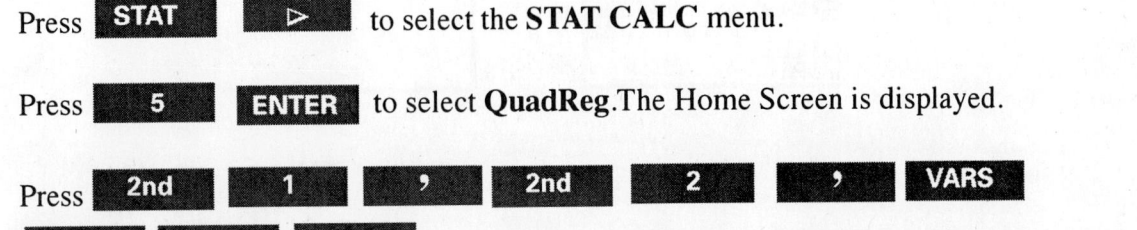

Press **STAT** **▷** to select the **STAT CALC** menu.

Press **5** **ENTER** to select **QuadReg**. The Home Screen is displayed.

Press **2nd** **1** **,** **2nd** **2** **,** **VARS** **▷** **1** **ENTER** to enter the names of the two lists, separated by a comma and the Y variable to which the function is to be pasted. The best–fit quadratic model is displayed on the home screen and pasted in the Y= menu.

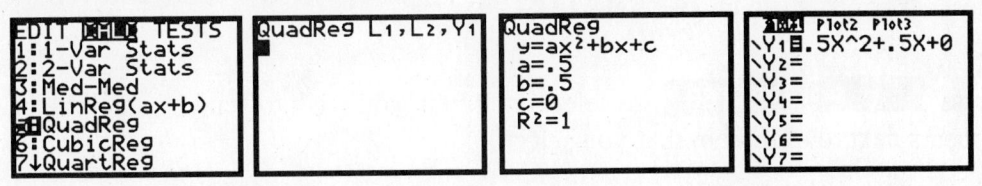

The value of a is the coefficient of the quadratic term. The value of b is the coefficient of the linear term and the value of c is the vertical intercept and constant term.

Graph the regression model.

Press **GRAPH** to display the graph of the regression model.

Adjust the viewing window if necessary. Use the table generated by the equation entered in the Y = menu to view additional values of the sequence.

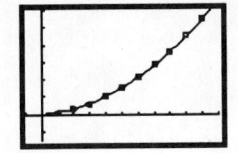

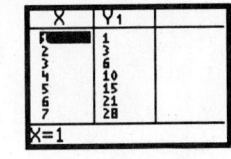

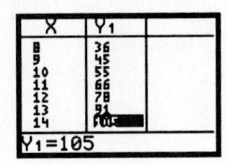

Transferring data between calculators: using LINK

Connect the two calculators using the connection cable.

Person receiving the data:

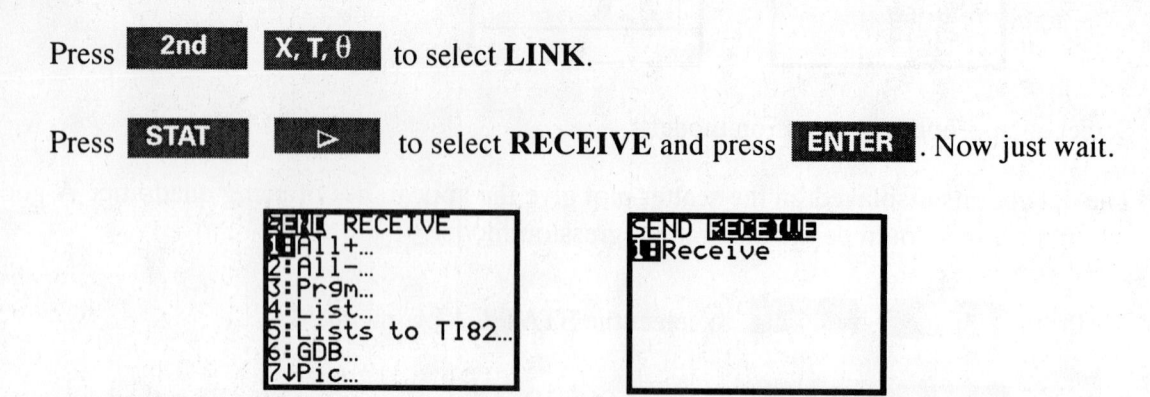

Press `2nd` `X, T, θ` to select **LINK**.

Press `STAT` `▷` to select **RECEIVE** and press `ENTER`. Now just wait.

Person sending the data:

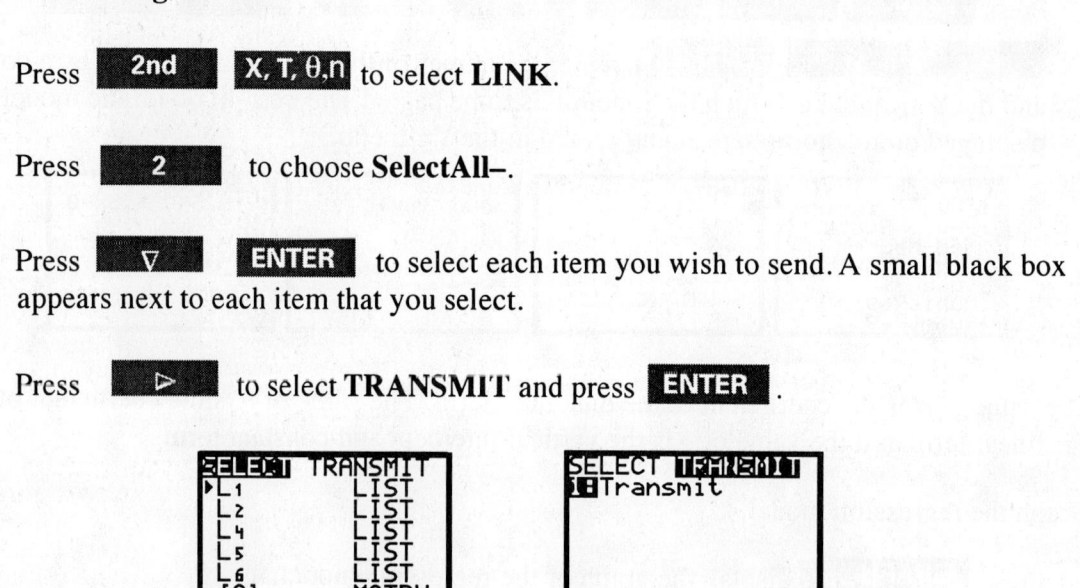

Press `2nd` `X, T, θ,n` to select **LINK**.

Press `2` to choose **SelectAll–**.

Press `▽` `ENTER` to select each item you wish to send. A small black box appears next to each item that you select.

Press `▷` to select **TRANSMIT** and press `ENTER`.

Disconnect calculators.

Entering and displaying a matrix

A matrix is a two-dimensional array of rows and columns. The dimensions of a matrix, the number of rows and the number of columns, is designated with the row number preceding the column number. The TI-83 stores as many as ten matrices using the matrix variables [A], [B], [C], [D], [E], [F], [G], [H], [I] and [J]. You can display, enter, or edit a matrix in the matrix editor.

Example: Enter the 3 x 3 matrix $\begin{bmatrix} 2 & -5 & 3 \\ 0 & 1 & -2 \\ -4 & 7 & 6 \end{bmatrix}$.

Set the dimension of the matrix. If you are using matrices to solve a system of equations, the number of equations is the row number and the number of variables is the column number.

Press **MATRIX** and arrow to **EDIT**.

Type **1** to select matrix [A].

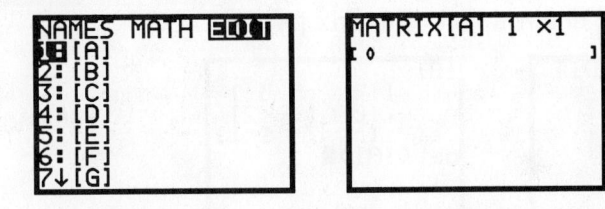

Enter the size of the matrix (in this case, 3 x 3) followed by the matrix elements.

Press **2nd** **MODE** to return to the Home Screen.

Press **MATRIX** **1** **ENTER** to display matrix [A] in the Home Screen.

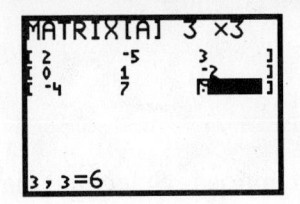

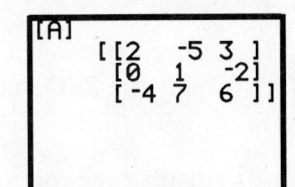

Calculating the determinant of a matrix

Example: Find the determinant of the 3 x 3 matrix $\begin{bmatrix} 2 & -5 & 3 \\ 0 & 1 & -2 \\ -4 & 7 & 6 \end{bmatrix}$.

Enter the matrix, if you haven't done so, as matrix [A].

Return to the Home Screen.

Press **MATRIX** and arrow to **MATH**.

Type **1** to select **det**, which represents a determinant.

The calculator returns to the Home Screen and displays **det**.

Press **MATRIX** **1** to display [A] after **det**.

Press **ENTER** to see the determinant of matrix [A].

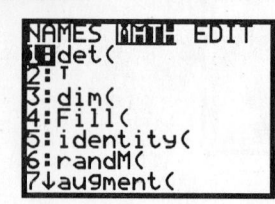

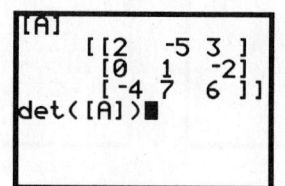

Deleting a matrix from memory

Press **2nd** **+** to access the **MEM** menu.

Press **2** to display the 2: Delete options.

Press **5** to display the matrix deletion options. Select the matrix you wish to delete.

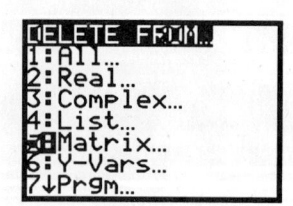

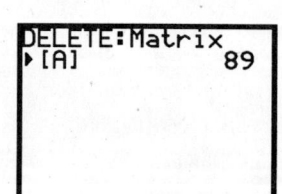

Press **ENTER** to delete the selected matrix.

Finding the inverse of a square matrix

Example: Find the inverse of the 3 x 3 matrix $\begin{bmatrix} 2 & -5 & 3 \\ 0 & 1 & -2 \\ -4 & 7 & 6 \end{bmatrix}$.

Enter the matrix, if you haven't done so, as matrix [A].

Return to the Home Screen.

Press **MATRIX** **1** **X⁻¹** **ENTER** to display the inverse of matrix [A] in decimal form.

The left and right arrow keys can be used to see all elements of the 3 x 3 inverse matrix.

Press **MATH** **1** **ENTER** to display the inverse matrix in fractional form.

273

Solving linear systems using matrix inverses

A system of linear equations can be solved by entering the coefficients as elements in an augmented matrix, then finding the solution using matrix row operations. If the coefficient matrix is square (same number of rows and columns) and the determinant is not equal to zero, you may solve the system by finding the product of the inverse of the coefficient matrix and the constant matrix.

Example: Solve the system
$$\begin{aligned} 2x - 5y + 3z &= 7 \\ y - 2z &= -3 \\ -4x + 7y + 6z &= 1 \end{aligned}$$

The coefficient matrix is $[A] = \begin{bmatrix} 2 & -5 & 3 \\ 0 & 1 & -2 \\ -4 & 7 & 6 \end{bmatrix}$ and the constant matrix is $[C] = \begin{bmatrix} 7 \\ -3 \\ 1 \end{bmatrix}$.

Enter the coefficient matrix as matrix[A], if you haven't done so.

Press **MATRIX** and arrow to **EDIT**.

Type **3** to select matrix [C].

Enter the size of the matrix (in this case, 3 x 1) followed by the matrix elements.

Press **2nd** **MODE** to return to the Home Screen.

The determinant of [A] is 12. The system has one solution since the determinant of the coefficient matrix is not zero.

If the linear system has a unique solution, the computation $[A]^{-1}[C]$ will return a column matrix that contains the solution to the system.

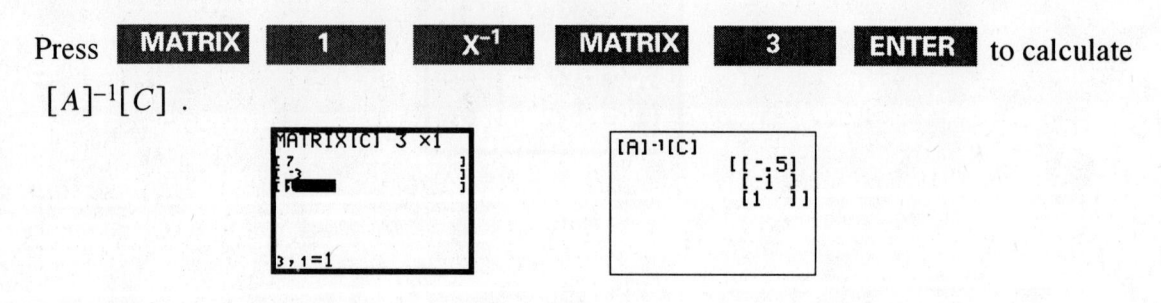

Press **MATRIX** **1** **X⁻¹** **MATRIX** **3** **ENTER** to calculate $[A]^{-1}[C]$.

The 3 x 1 solution matrix displays the solution to the system
$$\begin{aligned} 2x - 5y + 3z &= 7 \\ y - 2z &= -3 \\ -4x + 7y + 6z &= 1 \end{aligned}$$

Reading down the matrix, the solution is $x = -0.5$, $y = -1$, and $z = 1$.

Index of TI-83 Procedures

A

Access functions using the 2nd key 171
Add a random number to each element of a list 201
Adjust the view window 180
Adjust the display contrast 167
Apply arithmetic operations to an existing list to create new list 197
Arithmetic computations 172
ASK mode 177

C

CALC 190
Calculate the determinant of a matrix 214
Calculations 172
Change the mode 167
Change numbers in an existing list 196
Change the view window 184
Clear data from a list 195
Connected Mode for graphs 188
Construct a scatter plot of data 203
Construct a line plot of data 204
Continuous function graph 188
Control the TRACE option 186
Create a linear regression model 209
Create a quadratic regression model 211

D

Darken or lighten screen 167
Default view window settings 179
Delete a matrix from memory 215
Determinant of a matrix 214

Discrete graph display 188
Display coordinates of points on a graph 185
Display a discrete graph 188
Display a table of values 177
Dot Mode for graphs 188

E

Edit a previously-entered expression 173
Enter and display a matrix 214
Enter data into a list 195
Enter data into a list from the Home Screen 196
Enter expressions in the Y = menu 176
Evaluate a function 179, 180
Exponentiation 172

F

Factorial function 176
Find zeros of a function 191
Find intersection points 192
Find a function's maximum 193
Find a function's minimum 194
Finite differences 205
Finite ratios 205
Friendly view windows 186
Function evaluation: storing input 179
Function evaluation using Ask mode 181
Function evaluation using function notation 180

G

Generate data using an existing list 202

Generate data from the Home Screen 198
Generate new data on the list screen 197
Generate a list of data using the sequence function 199
Generate a list of random numbers 200
Graph a discrete function 203
Graph a function 182
Graph the regression model 210
Graph a scatter plot of data 203
Graph the solution interval to inequalities in one variable 190
Graphics screen 169

H

Home screen 168

I

Important keys and their function 170
Inequalities: graphing a solution interval 190
Insert new numbers into an existing list 196
INT function 174
Intersection points: finding 192
Inverse of a Square Matrix 215

K

Keyboard organization 166

L

Line plot of data 203
Linear regression models 209
List screen 169
Lists of data 195

M

Maximum of a function 193
Matrix inverses 216
Menus 168
Menu screens 169
Minimum of a function 194
MODE menu 188

N

Negative numbers 172

O

Obtain fractional results 174

P

Parametric graphing 189

Q

Quadratic Regression models 211

R

Raise a number to a power 172
RAND function 175
Reciprocal function 173
Regression equation 207
Regression models: graphing 210
Regression models: linear 209
Regression models: quadratic 211

S

Scatter plot of data 202
Scientific notation (3.46E11means) 173
Select a menu item 168

Setting a view window 183
Standard viewing window 183
STAT PLOT 204
Store a function input 179
Store the regression model 210
Store a value for a variable 178
Store a newly-generated list 197

T

Table screen 169
TRACE 186
Transfer data between calculators 213
Turn on/off Stat Plots 204
Turn the calculator on/off 166

V

View a newly-generated list 197
View window adjustments 180
View window standard settings 179

W

Working With Lists 195

Z

ZOOM options to adjust the viewing window 186
Zeros of a function 188